SpringerBriefs in Electrical and Computer Engineering

T0205973

For further volumes:
http://www.springer.com/series/10059

Dubravko Ćulibrk · Dejan Vukobratovic
Vladan Minic · Marta Alonso Fernandez
Javier Alvarez Osuna · Vladimir Crnojevic

Sensing Technologies
For Precision Irrigation

 Springer

Dubravko Ćulibrk
Dejan Vukobratovic
Vladan Minic
Vladimir Crnojevic
Faculty of Technical Sciences
University of Novi Sad
Novi Sad
Serbia

Marta Alonso Fernandez
Javier Alvarez Osuna
IMAXDI Real Innovation
Vigo
Spain

ISSN 2191-8112   ISSN 2191-8120 (electronic)
ISBN 978-1-4614-8328-1 ISBN 978-1-4614-8329-8 (eBook)
DOI 10.1007/978-1-4614-8329-8
Springer New York Heidelberg Dordrecht London

Library of Congress Control Number: 2013943574

Printed on acid-free paper

Springer is part of Springer Science+Business Media (www.springer.com)

Preface

This text aims to obtain a snapshot of the situation in the multidisciplinary, vibrant, and rapidly evolving space of precision irrigation and obtain valuable insights with technologies that can be employed and leveraged in the pursuit of the development of effective precision irrigation systems.

This book provides an overview of state-of-the-art sensing technologies relevant to the problem of precision irrigation, an emerging field within the domain of precision agriculture. Applications of wireless sensor networks, satellite data and geographic information systems in the domain are covered in the text. In addition to discussing the basic concepts of the technologies surveyed, an emphasis is placed on the practical aspects that enable the implementation of intelligent irrigation systems using the technology commercially available. As such, we hope that it will be of use to the broader audience interested in this theme. The text is organized in five chapters, each concerned with specific technology form the diverse set of ICT used to address the problem of optimal crop irrigation.

Chapter 1 is concerned with the application of wireless sensor networks, which can be used to monitor the microclimatic environment at an unprecedented scale. The discussion within this chapter starts with the aspects of sensor node hardware and the solutions provided by different manufacturers. An overview communication protocols (both high- and low-level) is provided next. Finally, we take a look at the available sensor networks solutions designed specifically for smart irrigation.

Chapter 2 will provide an overview of remote sensing data that can be used to assist precision irrigation. Aspects of different satellite data that can be used are discussed: satellite missions designed to assist irrigation, different products provided, data format and, in particular, ways to access the data for use in your own system.

Chapter 3 deals with the aspects of Geographical Information Systems (GIS) that are an inherent part of any precision irrigation system, as the data collected and processed needs to be stored, processed, and visualized. We will provide overview of most-commonly-used GIS, both open-source and proprietary. GIS applications, designed specifically for precision agriculture, are discussed in some detail.

Chapter 4 is dedicated to concluding remarks and some best-practice recommendations.

Acknowledgments

The authors would like to thank the consortium of the FP7-Project ENORASIS for their kind support, and the valuable contributions of Jérôme Donnadille and Panagiotis Symeonidis.

Contents

Chapter 1
Wireless Sensor Network Technology for Precision Agriculture

Precision agriculture demands intensive field data acquisition. One of the keys to understanding productivity variability lays in frequent data acquisition and interpretation. Wireless sensor networks (WSN) are a relatively new and rapidly developing class of wireless communication networks which can provide processed real time field data from sensors distributed in the field. The sensor nodes deployed on the field measure various atmospheric and soil parameters. These measurements can help in making decision on irrigation (automating, semi automating), fertilizer and pesticide applications, intruder detection, pest detection, yield prediction, plant disease prediction, fire detection, etc. The first part of this brief is devoted to Wireless Sensor Network technology with particular focus on its application in precision agriculture.

Sensor Node Hardware

Development of sensor nodes and associated hardware is currently an active research area carried out in universities and companies around the world. The possibilities in this field are enormous because of the increasing need to look for new sensors for different applications, the advances in miniaturization, components to be integrated, or new features to save energy. The aim of this section is to present the sensor node hardware architecture of WSN nodes and to provide the preview of some typical characteristics of commercially available nodes.

Hardware Architecture

The design of the system architecture is crucial to the longevity of the sensor networks. Sample rates, precision, synchronization accuracy etc. are factors that also have a significant impact upon power consumption. The system architecture of a typical WSN node is depicted in Fig. 1.1. The node, usually called the mote, is comprised of six architectural components:

D. Ćulibrk et al., *Sensing Technologies For Precision Irrigation*,
SpringerBriefs in Electrical and Computer Engineering,
DOI: 10.1007/978-1-4614-8329-8_1, © The Author(s) 2014

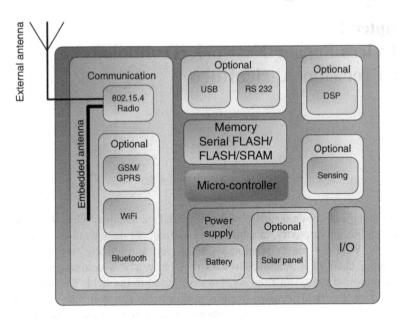

Fig. 1.1 Typical sensor mote architecture

1. Micro-controller unit (MCU)—is the central computational and control entity of a mote. The main tasks of the micro-controller are to manage the work of all components, to process data measurements and to communicate and store the data samples in the memory,
2. Memory modules—provide necessary memory capacity for mote operation including the space required for operation system, software applications and data storage,
3. Power supply component—which can rely on various forms of supply such as standard power network based power supply, energy harvesting options, or, most frequently, use of batteries,
4. Input/Output component—for external sensors and/or actuators connections and for node programming and debugging,
5. Radio communications module—for communication between nodes and between the node and the sink, and
6. Antennas—embedded or external.

Besides these essential modules, mote may have the following external components: (1) sensing component—sensors integrated on the node, (2) USB module, (3) RS-232 module,—both for communication with PC (programming or data reading), (4) solar panel—and/or any other energy harvesting component, (5) DSP—for signal processing (e.g., from multimedia sensors) (6) GSM/GPRS module, (7) Wi-Fi module, and (8) Bluetooth module—for data sending and communication with other devices.

Processing Power and Memory

Microcontroller represents the low-capacity processor which usually operates at low processor clock rates using computing architecture that ranges from 4 to 32-bit words. The central processing unit (CPU) is usually based on Reduced Instruction Set Computing (RISC) architecture with clock rates of several megahertz (MHz). It contains read-only (ROM) and random access (RAM) memory elements of several tens or hundreds of kilobytes (KB). Different "clocks" are used in order to provide local synchronization and Analog/Digital Converter (ADC) functionality. A low-energy operation mode and ability to wake power up, and return to sleep mode efficiently is a key factor in extending the lifetime of the network. Depending on the specific processor, power consumption in sleep mode ranges between 1 and 50 μW, while in active mode, it changes from 8 to 500 mW [1]. It should be noted that the trade-off between processing speed and power consumption is a key concern when selecting a processor for WSNs because higher processing power means higher energy consumption. The trend in the manufacture of microcontrollers for WSNs includes reducing power consumption while keeping or increasing their speed.

Memory is an external chip device with higher capacity than the internal ROM/ RAM memories, designed to store temporal data provided by different sources, e.g., sensors, network, etc. It should be noted that the flash writing/erasing is expensive, unlike the reading. Writing/reading ratio of energy consumption is about 80:1. Typical storage capacity of external chips ranges from 256 to 512 KB to several MBs or in extreme cases GBs.

Radio Circuitry and Antennas

The radio device provides wireless communication to the mote and supports the WSN-specific communication requirements such as low energy, sleep state, data rate and short distances. To extend battery lifetime, motes periodically wake up to acquire and transmit data by powering on the radio on and, upon completed communication task, powering it back off to conserve energy. The WSN radio must efficiently transmit signals and allow the system to go back to sleep mode with minimal power use. The main radio chips characteristics are:

- Capabilities: interface-type (bit, byte, packet level), data rates, distance range, frequency range (typical values include 315, 433, 868, 915 MHz and 2.4 GHz, ISM (Industrial, Scientific and Medical) band, number of channels, etc.
- Energy characteristics: power consumption to send/receive data, power efficiency, transmission power control, time and energy consumption to change states, etc.
- Radio performance: modulation, voltage range, gain (signal amplification), noise figure, receiver sensitivity, blocking performance (achieved BER in presence of

Table 1.1 Omnidirectional versus directional antennas

	Omnidirectional	Directional
Energy	More	Less
Throughput	More	Less
Collisions	More	Less
Connectivity	Stable	Intermittent
Discovery	Easy	Difficult
Coverage	Stable	Intermittent
Routing stretch factor	Less	More
Security	Less	More

frequency-offset interferer), out of band emissions, carrier sensing, Received Signal Strength Indicator (RSSI) characteristics, frequency stability (e.g., towards temperature changes), etc.

Antennas for WSN applications need to satisfy a number of additional properties, besides the standard ones such as the antenna efficiency. Low cost of the antenna represents the fundamental constraint. That is the one of the reasons why some nodes have Printed Circuit Board (PCB) antennas. WSN motes are (relatively) small, which means that antennas must also be small. It is reasonable to assume that the antenna can be committed to narrow-band at one of the most-frequently used ISM-frequencies 434, 868 MHz or 2.4 GHz.

Omnidirectional antennas are usually used in WSN, but directional antennas can also be used, e.g., when network topology is predefined. Directional antennas can focus their transmission energy in a certain direction. This feature gives rise to lower cross-interference and greater communication distance, given certain amount of energy. Directional antennas also can increase spatial reuse and reduce packet collisions and negative effects such as deafness. Table 1.1 shows advantages and disadvantages of both types of antennas.

Power Supply Sources

Even in cases where electric power might be available, the cost and difficulty of wiring the WSN motes remotely to the existing power mains is usually assumed to be prohibitive. Because of that, the main goal for the power supply source is to provide as much energy as possible at smallest cost, volume, weight, recharge time and longevity.

Batteries: Electrochemical batteries are the dominant source of electrical power for portable electronic devices, so they are logical choice for compact WSN nodes. Batteries are usually described as either *primary* (disposable) or *secondary* (rechargeable). Primary batteries are designed to be used only once because they use up their chemicals in an effectively irreversible reaction. Secondary batteries use chemical reactions that are reversible so they can be recharged by running a charging current in the opposite direction through the battery. Electrochemical batteries offer a relatively high energy density at low cost with no moving parts. The most important factor that affects battery lifetime is the discharge rate or the amount of current drawn from the battery. To avoid battery life degradation, the amount of current drawn from the battery should be kept under tight control [2].

There are four rechargeable battery technologies mainly used: Lead-Acid, Ni-Cad, Nickel-Metal Hydride (NiMH) and Li-Ion. As with every technology, each has advantages and disadvantages. Lead-Acid and Ni-Cad are reasonably priced, easily available and time-tested but are becoming increasingly unattractive because of environmental problems with their Lead and Cadmium content. That leaves Nickel-Metal Hydride and Li-Ion. Of those two, Li-Ion has the major advantage of operating at 3.6 V versus 1.2 V for the NiMH. Usually, one Li-Ion battery will suffice to directly power most WSN electronics whereas it would take 3 of the NiMH in series to do the same. The usage of secondary batteries is usually combined with some form of energy harvesting.

WSN motes on market are usually equipped with alkaline type of batteries e.g., AA (1.5 V, 1,800–2,600 mAh), CR123A (3 V, 1,500 mAh), C (1.5 V, 8,000 mAh) or with their rechargeable pairwise NiMH AA (1.2 V, 800–2,700 mAh), NiMH C batteries (1.2 V, 6,000 mAh) etc. For example, a node supplied with 2 AA batteries can last approximately 5 days if it is active all the time (20 mA consumption) and up to several years with power down (10 μA).

Energy harvesting: Portable energy reservoirs like batteries usually experience current leakages that drain the resource even when they are not used. A WSN that is dependent only on batteries has finite lifetime. Energy harvesting holds great promise for solving the problem of limited battery life and subsequent replacement especially where batteries are hard (or impossible) to replace/recharge. The harvester can be used as a lifetime extender for primary batteries or to recharge secondary batteries, because of its unlimited energy. Many interesting techniques to harvest ambient energy have been suggested over the past few years such as temperature differential, vibration, light and RF energy. Table 1.2 describes amount of energy that can be collected from different sources.

The harvesting circuits must have some type of energy storage to maintain power (e.g., throughout the night and during the cloudy days for solar-based sources). There are two options for power storage: rechargeable (secondary) batteries and supercapacitors.

Supercapacitor, or simply supercaps (also known as *ultracapacitor* or *double-layer capacitor*), are very large value capacitors with capacity up to 50 F. Working voltages are usually limited to 2.8 V so supercaps are stacked in series to handle the voltages needed to power the WSN device circuitry (3.3–5.0 V). Most important advantage of supercaps is the number of charge cycles. The energy storage device in a solar powered WSN will be charged and discharged daily (or more often) because of day/night cycles. Three years of operation will result in over 1,000 charge/discharge cycles. This can cause problems for many battery families because they are limited to 500–1,000 charge/discharge cycles before significantly reducing their capacity. Supercaps have no such limitation and can be

Table 1.2 Energy harvesting sources

Energy source	Energy density
Solar (outdoors)	100 mW/cm^2 (direct sun)
	0.15 mW/cm^2 (cloudy day)
Solar (indoors)	6 mW/cm^2 (standard office desk)
	0.57 mW/cm^2 (<60 W desk lamp)
Ambient radio frequency	1 μW/cm^2 (Wi-Fi)
	0.1 μW/cm^2 (GSM)
Temperature variation	4 μW/cm^3 (human motion—Hz)
	800 μW/cm^3 (machines—kHz)
Vibrations	500 μW/cm^2 (piezoelectric)
	4.0 μW/cm^2 (electromagnetic)
	3.8 μW/cm^2 (electrostatic)
Airflow	1 μW/cm^2
Acoustic noise	0.003 μW/cm^3 @75 dB
	0.96 μW/cm^3 @100 dB
Passive human-powered systems	330 μW/cm^2 (shoe inserts)
	7 W/cm^2 (heel strike)
	30 W/kg (hand generators)

charged/discharged millions of times. Table 1.3 describes comparison between supercaps and Li-ion battery.

Solar panels: Solar Photovoltaic (PV) panels are the most adequate for energy harvesting usage due to the energy amount than can be collected for the price of the panel. From an electrical point of view all PV cells look similar. They are a light-controlled current sources in parallel with a diode. Output current is a function of the physical size and efficiency of the cell. Output voltage per cell is essentially one diode drop; roughly 0.5–0.6 V. Higher voltage cells are composed of stacks of series cells [3].

As the prices of these solar cells continues to fall, the ability to put basic energy harvesting technology into everyday WSNs becomes realistic. There are two types of solar panels rigid and flexible—Fig. 1.2. Rigid solar cells are instantly recognizable as these are used in most portable equipment. These cells are moderately to expensively priced and provide light to electrical power efficiencies of 10–20 %. Flexible solar cells are much less efficient than the previous one but are usually

Table 1.3 Supercapacitor and Li-ion performance comparison

Function	Supercapacitor	Lithium-ion (general)
Charge time	1–10 s	10–60 min
Cycle life	1 million	500 and higher
Cell voltage	2.3–2.75 V	3.6–3.7 V
Specific energy (Wh/kg)	5 (typical)	100–200
Specific power (W/kg)	Up to 10,000	1,000–3,000
Cost per Wh	$20(typical)	$2 (typical)
Service life (in vehicle)	10–15 years	5–10 years
Charge temperature	−40–65 °C	0–45 °C
Discharge temperature	−40–65 °C	−20–60 °C

Fig. 1.2 Rigid end flexible
solar panel

cheaper and offer some mechanical flexibility. The latter feature can be very useful for applications requiring the cells to cover a curved surface. Efficiencies for these cells is in the 3–5 % range which is very low percentage of utilization but it can be a decent tradeoff versus the lower price. Typical solar panels available on the market provide 0.5–5 W, efficiency of ~ 15 %, output voltage of 5.5 V and are of dimensions of 10–20 cm of width/length.

Sensors for Precision Agriculture

There are two types of sensors, according to their power consumption, passive and active. A passive sensor does not require an electrical power source to generate an output, while an active one does. Typical examples of passive sensors suitable for precision agriculture applications include: temperature, humidity, atmospheric pressure, anemometers, pluviometers, motion (e.g., accelerometers or gyroscopes), soil moisture, soil water content, leaf wetness, solar radiation, piezoelectric microphones, etc., while active sensors include: gas detectors, radar, ultrasonic sensors and video cameras. In order to produce healthy/organic food the following sensors can also be very useful for agriculture purposes: Carbon Monoxide—CO, Geiger tube $[\beta, \gamma]$ (Beta and Gamma), Carbon Dioxide—CO_2, Oxygen—O_2, Methane—CH_4, Hydrogen—H_2, Ammonia—NH_3, Isobutene—C_4H_{10}, Ethanol—CH_3CH_2OH, Toluene—$C_6H_5CH_3$, Hydrogen Sulfide—H_2S, Nitrogen Dioxide—NO_2, Ozone—O_3, Hydrocarbons—VOC, etc.

Typical look of sensor mote boards with various attached sensors for precision agriculture applications is illustrated in Fig. 1.3.

Fig. 1.3 Agriculture board, gases board, radiation board

Sensor Node Protocols: Lower Layers

Sensor node hardware itself is only a necessary but not sufficient condition for efficient exploitation of WSN technology. Equally important part is represented by communication protocols designed to handle complex interconnection of wireless sensor nodes and maintain reliable and flexible data delivery process from each sensor node to the sensor network gateway towards the external networks. Given that communication, i.e. packet transmissions and receptions, represent the most intensive process in terms of energy consumption that is performed within the sensor node, its efficient and careful design is the main priority for establishment of the WSN technology as a reliable solution for environmental and in particular agricultural monitoring applications.

Communication protocols running within the sensor nodes could be roughly divided into two classes: (1) lower-layer protocols and (2) upper-layer protocols [4, 5]. This division corresponds to rough outlook of usual OSI model architecture, where lower-layer protocols are typically responsible for the communication and channel access functionality across a single link, whereas upper-layer protocols solve the problem of end-to-end data delivery between application-layer processes across the entire network. In the world of WSNs, the division between the lower and upper layer protocols is further accentuated due to the fact that the two protocol classes are developed and standardized by different standardization bodies. In particular, lower-layer protocols are part of responsibility of IEEE, more precisely, its 802.15.4^{TM} working group devoted exclusively to development of solutions for low-power wireless personal area networking solutions [6–10]. On the other hand, upper layer protocols typically build upon the IEEE 802.15.4^{TM} solution with different possibilities for provision of seamless and efficient IP connectivity of WSN nodes into the global Internet infrastructure. With IP connectivity, WSN nodes represent major components of the emerging framework of Internet of Things (IoT). However, unlike lower-layer protocols for which a single, IEEE proposed solution is widely accepted, for upper-layer protocols there are several possibilities such as ZigBee, 6LoWPAN and many others protocol stack solutions.

In this section, we provide a detailed description of the lower-layer part of the typical WSN protocol stack which is usually represented by one of the versions of the IEEE 802.15.4^{TM} standard. In the following, we will provide evolution of IEEE 802.15.4^{TM} standards on both OSI layers that comprise the lower-layer WSN protocols: the physical (PHY) layer and the medium access control (MAC) layer. The work on lowermost protocol layers: PHY and MAC layer, for WSN applications is almost exclusively the result of IEEE 802.15.4^{TM} working group effort (Fig. 1.18). The overall goal of IEEE 802.15.4^{TM} standards is to provide the Wireless Personal Area Networking (WPAN) solution for simple, low-cost and low data rate connectivity among widely spread devices, typically autonomous and without direct involvement of end-users. This kind of description includes different forms of sensor nodes in WSN, thus most of the available commercial WSN solutions use IEEE 802.15.4^{TM} standards as a basis for their MAC/PHY realization.

In the following, we provide details on IEEE 802.15.4™ PHY/MAC proposals, starting from the initial IEEE 802.15.4-2003™ standard and describing its extensions that were published over time [6–10].

IEEE 802.15.4™ Physical Layer

General Overview: Over the years, several IEEE 802.15.4™ PHY layers solutions are proposed and standardized, typically differing in operating frequency band, transmission range and data rates [11]. Different PHY technologies are investigated as part of different sub-working groups within IEEE 802.15.4™ working group. The core working group produced the first standard in 2003 known as the IEEE 802.15.4-2003 standard [6]. Due to ambiguities in the standard and driven by the need to reduce complexity of the initial solution, a special sub-working group has been established in 2006 with the goal of extending the initial standard under the name IEEE 802.15.4-2006, also known as IEEE 802.15.4b [7]. At about the same time, a parallel sub-working group defined novel PHY solutions capable of supporting ranging and localization functionalities, which are standardized within IEEE 802.15.4-2007, also known as IEEE 802.15.4a [8]. Further extensions that came after 2007 produced additional IEEE 802.15.4x, where x = {c, d} standards are already published and are mainly addressing novel PHY solutions for specific spectrum opportunities and system requirements in China (IEEE 802.15.4c) [9] and Japan (IEEE 802.15.4d) [10], while x = {e, f, g, j, k, m} standards are under development and address different application scenarios rather than PHY upgrades.

The main task of the 802.15.4™ PHY layer is to provide data transmission and reception service to the 802.15.4™ MAC layer, i.e. to provide for efficient MAC frame (MAC PDU) exchange among low-power sensor nodes. In technical terms, the main tasks performed by the PHY layer are [6]:

- PHY PDU encapsulated MAC PDU transmission and reception
- Channel frequency selection
- Clear Channel Assessment (CCA) signal generation for CSMA/CA MAC layer mechanism
- Link Quality Indicator (LQI) signal generation for received MAC PDUs
- Energy Detection (ED) signal generation for any of the underlying frequency channels
- Activation and deactivation of the radio-transceiver (Fig. 1.4)

Device Types and Network Topology: The IEEE 802.15.4™ standard defines two types of network devices: Full-Function Device (FFD) and Reduced-Function Device (RFD). The difference between the two is in the functionality they offer and the role in the network they are able to serve. FFD has more computing power, larger memory storage, sustain longer battery power supply, etc. Thus FFD is capable of serving any of the three network roles:

Fig. 1.4 IEEE 802.15.4™
lower-layer solution

1. PAN coordinator,
2. coordinator, or
3. network device.

In contrast, RFD offers only a minimal set of functions defined by the standard and is able to perform only a role of network device. By the standard–defined constraints, RFD can only associate to a single FFD at a time so the communication between RFDs can only proceed if the intermediate FFD is involved.

Using FFD and RFD devices, two different topologies are envisaged by the standard (Fig. 1.5: IEEE 802.15.4™ devices and topologies):

1. star topology, and
2. peer-to-peer topology.

Both topologies are introduced in the initial standard and preserved throughout standard upgrades. Also, both topologies require at least one FFD device included in the network as a PAN coordinator.

The star topology consists of a single PAN coordinator, a FFD device that can communicate directly with all other involved devices. Thus the PAN coordinator is the central network point which is responsible for establishing, organizing and maintaining all the network communication. Apart from the PAN coordinator, other nodes can be either FFD or RFD, however, they are all required to communicate only via the central PAN coordinator.

In contrast to the star topology, the peer-to-peer topology allows direct communication between the network devices. Thus multi-hop routing of network

Fig. 1.5 IEEE 802.15.4™
devices and topologies

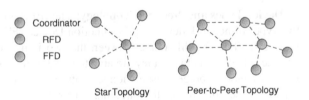

messages is independent of the PAN coordinator and is performed independently by each involved device. This flexibility allows for larger areas to be covered by the network with more complex topologies such as the mesh topology or so called cluster-tree topology. Similarly to the star topology, the peer-to-peer topology still requires at least a single FFD to serve as a PAN coordinator.

IEEE 802.15.4-2003 PHY Standard [12]: The design of the initial IEEE 802.15.4-2003 standard (the 2003 standard in the further text) has been mainly driven by the need for low complexity and low power consumption solution. Three different solutions have been developed, each one positioned one of the following three licence-free frequency bands, with the total of 27 user channels available for use (Fig. 1.6):

1. 868–868.6 MHz (1 channel) in Europe,
2. 902–928 MHz (10 channels) in North America, and
3. 2,400–2,483.5 MHz (16 channels) worldwide.

The 915 MHz band central frequencies are obtained as $f_c = 906 + 2(k - 1)$ MHz, where $k = \{1,2,...,10\}$, while for the 2.4 GHz band, the central frequencies are $f_c = 2,405 + 5(k - 11)$ MHz, for $k = \{11,12,...,26\}$, where k is the channel number.

The 868 MHz solution provides only a single channel and targets the data rate of 20 kbps. The 915 MHz band solution provides 10 channels on the 2 MHz spacing between neighboring channels and target data rate of 40 kbps. Finally, the 2.4 GHz solution provides 15 channels at the 5 MHz spacing aiming at the highest data rate of 250 kbps. All solutions use the Direct Sequence Spread Spectrum (DSSS) technique.

For all three solutions, the goal is to provide simple and energy-efficient transmission of PHY packets (PHY PDUs) over the wireless interface among sensor nodes. The structure and the size of the PHY PDU is illustrated in Fig. 1.7. The PHY PDU starts with 32-bit synchronization preamble required for the

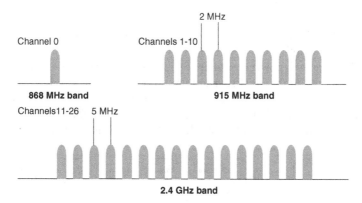

Fig. 1.6 IEEE 802.15.4™ frequency bands and channels

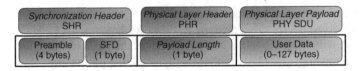

Fig. 1.7 IEEE 802.15.4™ PHY packet format

receiving node to lock the phase-lock loop circuits (PLL) to the incoming signal bit-rate. The preamble ends with 8-bit Start of Packet Delimiter (SPD). Finally, the last PHY PDU header field is 8-bit long indicator of PHY SDU (PHY packet content) length, which indicates the size of the remaining part of the PHY PDU which can be between 0 and 127 bytes.

In the 2.4 GHz solution, after the PHY PDU is formed and header fields completed, the transmission proceeds by applying 16-qary orthogonal modulation technique as follows. Firstly, PHY PDU bits are converted into symbols by forming groups of 4 consecutive bits from the PHY PDU content. Then each symbol is transformed into one of 16 orthogonal 32-bit chip pseudo-noise (PN) sequences defined in the standard. The resulting chip sequence is modulated onto the carrier using offset QPSK (O-QPSK) modulation. Clearly, this technique essentially represents DSSS resulting in the 8-fold signal bandwidth increase, as 32 bits are produced from each group of 4 bits. Thus the resulting chip rate from the 250 kbps input signal equals 2 Mcps (chips per second). The flow of PHY processing steps is illustrated in Fig. 1.8: IEEE 802.15.4™ PHY processing steps.

Both 868/915 MHz solutions use Binary Phase Shift Keying (BPSK) modulation (with differential encoding) in combination with 15-bit chip PN spreading sequence. The resulting chip rate results in 15-fold increase over the input bit rate and is equal to 300 and 600 Kcps.

The PHY design described above is mainly motivated by the needs for simplicity and low-power operation. In particular, for the most demanding 2.4 GHz solution, the solution based on orthogonal signaling trades bandwidth for improved sensitivity which is needed for low power operation. In addition, DSSS technique is preferable for short settling times of wide-band channel filters, which are needed for typically short active periods, i.e. extremely low duty cycles, of sensor nodes. Finally, constant-envelope modulations such as O-QPSK combined with half-sine shaped signals simplify the design and reduce the active power requirements leading to decreased power consumption. Overall characteristics of the PHY solutions proposed in the 2003 standard are presented in Table 1.4.

Apart from the PHY PDU transmission, the PHY layer is responsible for generating several PHY signals which are of immediate use for both the PHY and the MAC layer within the IEEE 802.15.4™ stack. This PHY functionality is

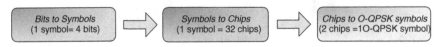

Fig. 1.8 IEEE 802.15.4™ PHY processing steps

Table 1.4 IEEE.802.15.4™-2003 PHY solutions overview

PHY solution	Frequency (MHz)	Spreading rate (kcps)	Modulation	Data rate (kbps)	Symbol rate (ksymbps)
868 MHz	868	300	BPSK	20	20
915 MHz	902–928	600	BPSK	40	40
2.4 GHz	2,400–2,483.5	2000	O-QPSK	250	62.5

usually referred to as the channel assessment functions. The signals generated at the PHY resulting from channel assessment include RSSI, CCA and LQI signals described below [13].

At the connector between the Radio Frequency (RF) circuitry and the antenna, the RF power measurements are made. For the transmitting signal, the Effective Isotropic Radiated Power (EIRP) is measured this way, while for received signals, these measurements are used to derive Received Signal Strength Indicator (RSSI). Clear Channel Assessment (CCA) signal is important for CSMA/CA functionality within the MAC layer. It is assessed by the PHY layer measurements in one of three modes:

1. Energy Detection (ED) threshold mode, where CCA delivers busy channel notification if the surrounding signal energy detected is above some predefined energy threshold,
2. Carrier Sense (CS) mode, where the PHY examines if there are IEEE 802.15.4™ compliant signals in the surrounding irrespective of the energy level of the detected signal, and
3. CS with ED mode, which combines the previous two modes. Finally, Link Quality Indicator (LQI) returns to the sender information on quality of the packet transmission. It is reported on a per PHY PDU basis and is usually derived from ED signal or as a separate Signal to Noise Ratio (SNR) estimation.

IEEE 802.15.4-2006 (IEEE 802.15.4b) PHY Standard [14]: The first standard update in 2006, known as IEEE 802.15.4b, introduced further upgrades for PHY layer solutions in 868 MHz and 915 MHz band. The goal was increasing the data rate in these bands from 20 and 40 kbps to 250 kbps, the same data rate achievable by the 2.4 MHz solution. The 2.4 GHz band was far more widely adopted initially due to worldwide availability of this band, higher data rate and more available channels, lower power consumption (shorter transmission/reception times due to higher data rates), RF band which is mature and well exploited by other technologies (e.g., IEEE 802.11™ or Bluetooth).

On the other hand, the 868/915 MHz band offers advantages in terms of favorable propagation performance (such as less absorption and reflection) resulting in larger communication ranges. Thus it was of interest, especially for outdoor applications, to increase the rate of the 868/915 MHz band to the high rate performance of 250 kbps.

The data rate improvement is achieved with additional complexity costs by applying two different approaches. The first approach changed the modulation

Table 1.5 IEEE.802.15.4-2006 PHY solutions overview

PHY solution	Frequency (MHz)	Spreading rate (kcps)	Modulation	Data rate (kbps)	Symbol rate (ksymbps)
868 MHz	868	400	O-QPSK	100	25
868 MHz	868	400	ASK	250	12.5
915 MHz	902–928	1,000	O-QPSK	250	62.5
915 MHz	902–928	1,600	ASK	250	50

scheme in 868/915 MHz band from BPSK to O-QPSK scheme, the one used in 2.4 MHz band, resulting in data rates of 100 and 250 kbps in 868 and 915 MHz band, respectively. The second approach was based on so called Parallel Sequence Spread Spectrum (PSSS) technique, which uses BPSK and Amplitude Shift Keying (ASK) to provide 250 kbps for both 868 and 915 MHz solutions. The characteristics of the novel PHY schemes included in the 2006 standard are presented in Table 1.5 which extends the information provided in Table 1.4.

To identify which PHY scheme is applied in which channel, so called Channel Page numbers are introduced (Table 1.5). The Page 0 is defined by the initial set of 27 channels introduced for 2003 standard. The 2006 standard introduced two new channel pages, Page 1 and Page 2, over the same set of frequency bands and corresponding channels as in the 2003 standard (i.e. 868/915/2,400 MHz bands). Both Pages 1 and 2 use only 868/915 MHz channels (channels 0–10) where Page 1 refers to the first, O-QPSK approach, and Page 2 refers to the second PSSS approach. Pages 3–6 were introduced and defined for later releases of standards described in the sequel.

IEEE 802.15.4-2007 (IEEE 802.15.4a) PHY Standard [15]: The main motivation behind the new standard release in 2007, the IEEE 802.15.4a standard, was in introduction of precision ranging functionality in sensor nodes [15, 16]. It was predicted that sensor nodes with precision ranging functionality could significantly improve the scope of possible applications of WSN technology. However, the precision ranging functionality is hard to implement with low signal bandwidths of 2003 and 2006 standard solutions. Thus an alternative PHY solution was needed complemented with amendments to the MAC layer functionality to support for high-precision distance measurements in WSN.

The sub-working group involved in the novel PHY layer design proposed two alternative solutions:

1. Ultra Wide Band (UWB) impulse radio, and
2. Chirp Spread Spectrum (CSS) solution.

The first solution operates in three different UWB bands within unlicensed UWB spectrum whereas the second solution operates in 2.4 GHz ISM band (Table 1.6).

The UWB solution brings into the WSN world a lot more than possibility of precision ranging. With increased bandwidth and UWB technology, larger ranges and significantly higher data rates are possible, opening the way for possible WSN

Table 1.6 IEEE.802.15.4-2007 PHY solutions overview

PHY solution	Frequency (MHz)	Spreading rate (kcps)	Modulation	Data rate (kbps)	Symbol rate (ksymbps)
UWB	250–750	Pulse repetition	BPSK/PPM	110	55
	3244–4742	frequency (PRF):		850	425
	5944–10234	3.9, 15.6 and 62.4 MHz		6810	3,405
				27240	13,620
CSS	2400–2483.5	Variable	D-QPSK	250	/
			8 and 64-ary	1,000	
			orthogonal codes		

applications requiring e.g., high-quality sound or video transmission. Three frequency bands are allocated for UWB transmission:

1. sub-GHz band (250–750 MHz),
2. low-band (3.244–4.742 GHz), and
3. high-band (5.944–10.234 GHz).

The sub-GHz band provides a single channel (channel 0), the low-band provides 5 channels (channels 1–5), and the high-band provides 10 channels (channels 6–15). All except three channels (channel 4, 7 and 15) are 500 MHz channels, while the remaining three are of larger bandwidth thus offering higher ranging precision then the standard 500 MHz channels (whose ranging precision is of the order of 1 m). Channels 0, 3 and 9 (one from each band) are mandatory and any 802.15.4™ a compliant device should be capable of communicating over these channels.

The PHY transmission process for UWB solution is based on the impulse radio technique, which transmits band-limited pulses whose bandwidth is matched to the channel bandwidth. A symbol of information consists of a sequence of pulses called burst, where depending on the number of pulses in the burst, different information symbol rates can be obtained. In a typical realization, two information bits are encoded into symbols using a combination of Pulse Position Modulation (PPM) and BPSK modulation, where the first bit determines the position of the burst while the second one determine its phase. The UWB solution provides four possible data rates: 0.11, 0.85, 6.81 and 27.24 Mbps, out of which only the 0.85 Mbps data rate is mandatory for standard compliant devices [16].

The CSS solution is developed for larger communication range and data delivery to high speed vehicular users' scenarios as the main target. The CSS transmit over 14 allocated 5 MHz spaced channels in the 2.4 GHz ISM band, positioned between 2.41 and 2.486 GHz. The supported data rates of the CSS solution are 250 kbps and 1 Mbps. CSS solution represents a spread sequence technique which is similar to both DSSS and UWB techniques. It is based on chirp signals which are time-limited (windowed) signals whose frequency increases or decreases quickly and in a linear fashion (i.e. the speed of frequency change is constant over time) over some frequency range during the short chirp-signal duration. In the CSS solution, a chirp symbol is built by combining smaller sub-chirps. The chirp modulation is further

Table 1.7 IEEE.802.15.4-2009 PHY solutions overview

PHY solution	Carrier frequency (MHz)	Spreading chip-rate (kcps)	Modulation	Data bit rate (kbps)	Symbol rate (ksymbps)
780 MHz O-QPSK	780	1,000	O-QPSK	250	62.5
780 MHz MPSK	780	1,000	MPSK	250	62.5
950 MHz DSSS	950	300	BPSK	20	20
950 MHz GFSK	950	900	GFSK	100	25

combined with differential QPSK (DQPSK) modulation and 8-ary or 64-ary bi-orthogonal coding resulting in the 1 Mbps and 250 kbps data rate solutions. Overall, the CSS solution results in a solution favorable for long-range and high-speed mobile users WSN communications [17].

IEEE 802.15.4-2009 (IEEE 802.15.4c/d) PHY Standard [18]: Two additional PHY layer amendments are published in 2009 standard, the IEEE 802.15.4c standard which targets usage of 780 MHz band available in China, and the IEEE 802.15.4d standard that supports 950 MHz available in Japan.

The IEEE 802.15.4c standard introduces two novel PHY solutions (Table 1.7):

1. O-QPSK based solution, and
2. MPSK based solution.

Both solutions are designed for 780 MHz band and can use one of eight available channels at a mutual distance of 2 MHz. The O-QPSK solution supports 250 kbps data rate using similar PHY layer processing as in original 2.4 MHz solution from the 2003 standard, except that the chip rate is reduced to 16-bit PN sequences per 4-bit information symbols. The MPSK solution also provides for 250 kbps of data rate, however, by employing phased encoding of 16-bit PN sequence and MPSK modulation onto carrier.

For usage in Japan, two PHY schemes are included in the IEEE 802.15.4d standard (Table 1.7):

1. DSSS/BPSK based solution, and
2. Gaussian Frequency Shift Keying (GFSK) based solution.

The first solution is almost identical to the 868/915 MHz band solution in the original 2003 standard. It provides 20 kbps data rate, however, in the channels within the 950 MHz band available in Japan. Operating in the same band, the second solution provides 100 kbps of data rate, where the data is first scrambled by use of 9-bit PN sequence and then modulated using GFSK modulation.

IEEE 802.15.4™ Medium Access Control Layer

General Overview: The IEEE 802.15.4™ MAC layer is developed with the main goal of coordinating access of WSN nodes to the physical medium [13, 19]. It is heavily inspired by the IEEE 802.11™ Wireless LAN (WLAN) MAC protocol, however, with additional simplifications introduced wherever possible. The main set of tasks the MAC layer in 802.15.4™ takes care of is listed as follows [18]:

- Generating network Beacons if the device is a Coordinator,
- Synchronization to the Beacon,
- Support for PAN association and de-association,
- Support for sensor node security,
- Application of CSMA/CA algorithm for PHY channel access,
- Handling and maintaining Guaranteed Time Slot (GTS) mechanism, and
- Providing a reliable MAC to MAC entity communication between devices.

The above tasks are described in the following text within the process of data communication between network devices using different possible channel access modes and, where applicable, applying CSMA/CA algorithm for random channel access.

MAC Frame Format: The data transmission among MAC entities of WSN devices assumes exchange of MAC PDUs, usually called MAC frames. The structure of a MAC frame is illustrated in Figure and, similarly to all data link layer protocol frames, it consists of three major parts:

1. MAC frame header (MHR),
2. MAC payload (MAC SDU, i.e. user data), and
3. MAC footer (MFR).

The above MAC PDU format is general and applies to all four types of MAC frames. These are:

1. Beacon frames,
2. Data frames,
3. Acknowledgement (ACK) frames, and
4. MAC Command frames.

The type of the MAC frame is indicated in the first field of the MHR, within the Frame Type subfield of the Frame Control field. Clearly, the most important frames are data frames which transmit upper layer messages, however, the remaining frame types are crucial for MAC protocol operation.

MHR part contains three header fields: Frame Control field, Sequence Number field, and Address Fields, whose details can be found in the standard. There are two options for device addressing within the network:

1. 64-bit IEEE (MAC) Address, and
2. 16-bit Short Address.

IEEE address is represented in standard 64-bits IEEE MAC layer format for LAN physical interfaces. It is assigned by device manufacturer according to global IEEE addressing rules and no two devices can have the same MAC address. In WSN applications, it is frequently called extended address. In contrast, the short address is 16-bit address which is local to the network, which means that two devices in different WSNs can have the same short MAC address without causing any problems to each other. The short address is usually allocated by a coordinator during the process of the device association. Not all frame types use all the address fields, e.g., beacon frames include only the source address field.

MHR payload field carries the user data of upper protocol layers that use the MAC layer service of data transmission over the physical channel. MAC payload size has to conform to the requirements of the PHY layer, which typically bound the whole MAC frame to a maximum of 127 bytes. Thus the maximum payload size depends on the MHR format applied (e.g., which type of addresses are used) and is selected so that the overall frame size is kept under the PHY constraints. Not all MAC frame types transmit the payload, e.g., ACK frames do not have the MAC payload field.

Finally, MFR represents a 16-bit frame-check sequence (FCS) based on the Cyclic Redundancy Check (CRC) polynomial $x^{16} + x^{12} + x^5 + 1$. FCS is calculated at the receiving device and if the CRC check is passed, the MAC frame content is processed, otherwise, the MAC frame is considered erroneous [20].

MAC Communication Modes: Communication within the 802.15.4™ network can be maintained using two different modes:

1. Beacon Enabled Mode (or Contention-Free mode), and
2. Beacon-Less Mode (or Contention mode).

These modes offer two different options for how much control is involved in communication process. In the first beacon enabled mode, a PAN coordinator takes the role of a central network point which synchronizes and coordinates all the device communications within the network. In the second beacon-less mode, there is no central control and the control of the channel access is distributed over all devices in the network by application of the celebrated Carrier-Sense Multiple Access with Collision Avoidance (CSMA/CA) algorithm. In the following, we provide details of the two communication modes since they underlie the main principles of how data transmission is performed with the WSN (Fig. 1.9).

In the beacon enabled mode, a FFD called PAN coordinator establishes time synchronization among all devices in the network by introducing superframe time structure. A superframe period is marked at the beginning by so called beacon frame (thus the name beacon enabled mode). Beacon frames are periodic and generated and broadcasted by PAN coordinator to all the devices in the network. Following the beacon frame, a superframe further consists of time slots that can be used by devices to communicate over network. Each beacon frame carries important information that synchronizes communication among devices in the

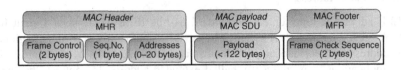

Fig. 1.9 IEEE 802.15.4™ MAC frame format

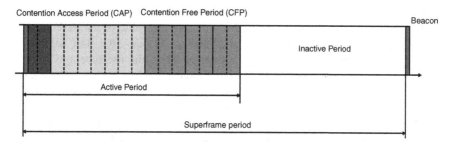

Fig. 1.10 IEEE 802.15.4^{TM} MAC superframe structure

upcoming superframe, including the format of the superframe. Every device in the network has to listen during the beacon frame in order to get information on the upcoming superframe structure.

In general, the superframe may be divided into two parts, an active period and an inactive period, as illustrated in Fig. 1.10. In the active period, after the beacon frame, 15 equal duration time slots follow that are further subdivided into Contention Access Period (CAP) and Contention Free Period (CFP). The CAP slots are mandatory while the CFP slots are optional and limited to up to seven time slots. The information on the number of time slots assigned to each CAP and CFP is contained within the beacon frame. During the CAP, devices use slotted CSMA/CA algorithm to contend to the shared medium. On the other hand, CFP period is divided into Guaranteed Time Slots (GTS) allocated in advance to those users that explicitly requested this kind of guaranteed service. GTS allocated to a single user may contain more than a single time slot during which it is guaranteed to the user that its communication will be serviced without competing users. After the CFP, there may be inactive period which lasts until the final time slot of a superframe. The total length of the superframe T_{SF} can be decided by the PAN coordinator and is of the form $T_{\text{SF}} = T_{\text{BSF}} * 2^{\text{BO}}$, where $T_{\text{BSF}} = 16$ is the baseline superframe duration, and BO, $0 \leq BS \leq 15$ is the beacon order. For $BS = 0$, there is no inactive period and for $BS = 15$, the mode is beacon-less. During the inactive part of the superframe, the nodes are powered down into the sleep mode to save energy. Thus the inactive mode is introduced to control the duty cycle of the WSN nodes.

In the beacon-free mode, there is no central coordination among devices and access to the physical channel is resolved by unslotted CSMA/CA algorithm deployed in each network device. The unslotted CSMA/CA (in the following text only CSMA/CA) does not put any constraints on so called back-off unit time intervals (to be described in the following paragraph) which are left to be selected freely and independently by participating devices, in contrast to the slotted CSMA/CA used in beacon enabled mode where this back-off unit time interval has to be aligned with the time slot duration. Due to its significance to WSN operation, in the following, we shortly describe the CSMA/CA algorithm.

The CSMA/CA algorithm proceeds as follows. Firstly, the variable counter of number of back-offs (NB) is set to $NB = 0$ and so called back-off exponent is set to its minimal value BE = minBE. The device wishing to transmit the data first checks the situation in the surrounding by sampling the value of the PHY signal CCA (Clear Channel Assignment). If the medium is busy, the user selects a random back-off period and waits until it expires. Then, the user attempts again, and repeats the attempts until the CCA signal indicates a free medium. Once this happens, the device transmits its message but the transmission needs not to be successful due to possible collision with other device that may have started transmitting almost simultaneously. If this happens, the device gets into the retransmission back-off algorithm whose unslotted version is depicted in Fig. 1.11.

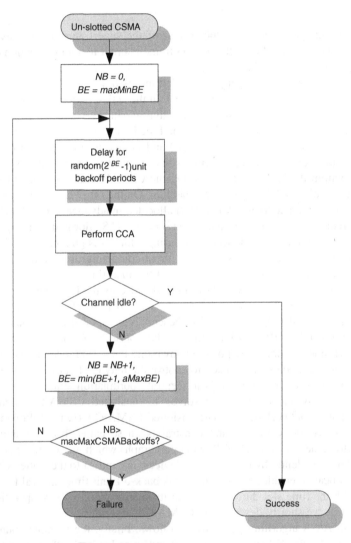

Fig. 1.11 IEEE 802.15.4^{TM} MAC unslotted CSMA/CA

During the back-off stage, the device selects a random waiting period from the interval $[0, 2^{BE} - 1]$. After this period expires, the device will perform CCA again, and if the channel is free, it will start transmission, otherwise, the variable counters NB and BE will be incremented. This procedure will be repeated until the successful transmission or until the maximum number of back-offs is reached when the data message is simply dropped.

IEEE Transmission Scenarios: Given the above described Beacon-Enabled and Beacon-Less modes, there are three major transmission scenarios defined within the IEEE 802.15.4™ standard: Coordinator to Device, Device to Coordinator and Device to Device (Peer to Peer).

Coordinator to Device: in the transmission scenario with the Beacon–Enabled mode in which the Coordinator has a data to transmit to a network Device, the Coordinator will first indicate its intention using the Beacon targeting the selected Device. As other devices, the targeted device listens to the Beacon and gets information on the upcoming data delivery. Upon processing the Beacon information, the Device will select a free slot and send a data request message to the Coordinator indicating that it is ready to receive the content within the specified time slot. When the Coordinator receives the data request message, it will perform data packet transmission in a specified time slot. After the data block is successfully received by the Device, it responds with an acknowledgement message to the Coordinator, as illustrated in Fig. 1.12a.

In the Beacon-Free mode, the situation is a bit more complicated, since there is no predefined moment when the Device listens to the Coordinator. Thus in this scenario, the Coordinator stores the data in its local memory and waits until the Device explicitly asks if there are any messages stored for its reception. When the Device

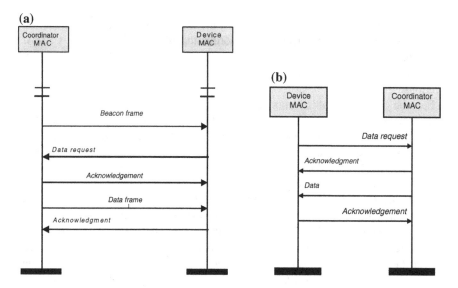

Fig. 1.12 Coordinator to device transmission. **a** Beacon-Enabled mode **b** Beacon-Free mode

Fig. 1.13 Device to
coordinator transmission

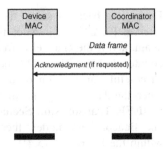

asks the Coordinator if there are any messages targeting the Device, the Coordinator checks its storage buffer, and if the answer is 'yes' it will send ACK message to the Device. Immediately after the ACK message, the Coordinator sends data to the Device which responses back with its own ACK message, as presented in Fig. 1.12b.

Device to Coordinator: In the Beacon-Enabled mode, if a device wants to transmit data to the Coordinator, the Device waits for a Beacon and then follow the time slot structure defined in the Beacon frame. More precisely, the Device will wait for the slots belonging to the Contention-Free Period (CFP) and attempt transmitting using slotted CSMA/CA protocol. Once the transmission is successful, the Device will receive ACK message from the Coordinator and the transmission is successful.

In the Beacon-Free mode, the Device has no reference signal to wait for and there is no predefined time slotted structure, so the Device starts its transmission immediately using the unslotted CSMA/CA access method. Once the transmission is successful, the Coordinator sends ACK message to acknowledge the reception to the Device. Figure 1.13 illustrates this situation, while the Beacon-Enabled case differs only in periodical Beacon frame transmitted by the Coordinator.

Device to Device: In Beacon-Free mode, the Device to Device communication proceeds similarly as in the Device to Coordinator mode using unslotted CSMA/CA algorithm. This poses a restriction on the devices in the network to be constantly active which makes this approach typically unfavorable for energy efficient solutions.

For the Beacon-Enabled mode, the Device to Device communication is somewhat complicated due to possibilities that devices are affected by beacon signals provided from different sources cause possible synchronization issues.

The above text describes the frame structure and basic channel access operations of the IEEE 802.15.4-2003 MAC protocol. Standard updates that followed the 2003 standard are shortly described below.

IEEE 802.15.4™ MAC Standard Updates: The standard updates that followed the 2003 standard [18] brought further improvements and additions to the original 2003 MAC standard [14, 15, 21, 22]. In the 2006 standard [14], a number of ambiguities are removed from the original standard, many procedures are simplified and clarified and the whole standard was made more accessible and implementation friendly. Among the changes, CSMA/CA procedure is improved, GTS time slots within the superframe are left optional for use and further enhancements in security mechanisms are introduced.

Due to significant changes in PHY layer technology in the 2007 standard [15], where UWB PHY solution is introduced, it was necessary to change the MAC layer functionality. In particular, UWB solution is not the standard PHY technology that applies carrier signals in a usual sense, thus CSMA/CA algorithm is not the most suitable one for this scenario. Instead, it has been shown that ALOHA medium access protocol works satisfactory and is able to achieve considerably larger throughput, which is the reason why ALOHA is included in the standard as a basis for a MAC solution in the 2007 UWB standard version. However, as the ALOHA solution works well only for small to medium network density, the CSMA/CA solution is still included as optional for dense network scenarios.

Finally, both 2009 versions of the standard [13, 21] do not effectively change any of the fundamental properties of the CSMA/CA based access scheme of the original 2003 standard.

Sensor Node Protocols: Upper Layers

There are, nowadays, two main specifications for implementing the communication upper layers in wireless sensor networks: ZigBee™ and 6LoWPAN.

ZigBee®

ZigBee® is a bidirectional wireless communication standard, developed by the ZigBee® Alliance [23] and designed for low-cost and low-power consumption communication. It is built on the foundation of the IEEE 802.15.4™ standard, which defines the two lower layers: the physical (PHY) layer and the medium access control (MAC) sub-layer. The ZigBee® Alliance provides the network (NWK) layer and the framework for the application layer, Fig. 1.14 represents the ZigBee® layers and its interactions.

In ZigBee® networks there are three types of devices as specified in the Ref. [24] coordinator, router and end device. A coordinator is responsible for initializing maintaining and controlling the network. The supported topologies are: star, tree and mesh topology, as shown in Fig. 1.15.

A star network has a coordinator to whom end devices are directly connected, while in tree and mesh networks devices communicate each other through multihop paths. The network contains one coordinator and multiple ZigBee® routers. An end device can join the network by associating directly with the coordinator or a router. These end devices are generally responsible for sensing.

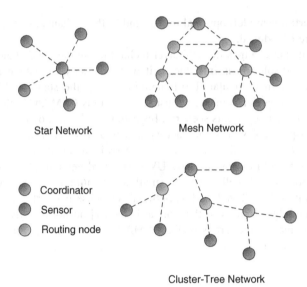

Star Network   Mesh Network

Coordinator

Sensor

Routing node

Cluster-Tree Network

Fig. 1.14 ZigBee® stack

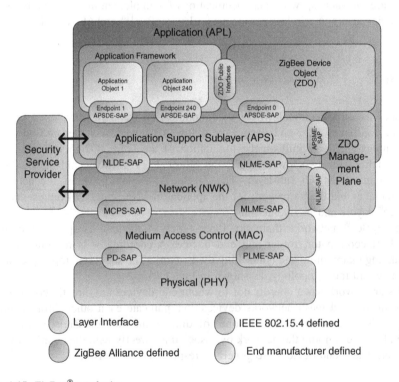

Fig. 1.15 ZigBee® topologies

ZigBee® Network Layer

Description

ZigBee® network layer includes functionality to ensure the correct operation of the IEEE 802.15.2™ MAC sub-layer and to provide a service interface to the application layer. To interface with the application layer, the network layer includes two service entities: the data service (NLDE) and the management service (NLME).

The services provided by these entities to the application layer are accessed through their associated Services Access Points (SAPs), acting as an interface between the application and the MAC sub-layer via the MCPS-SAP (Medium access control common part sub-layer SAP) and MLME-SAP (Medium access control sub-layer management entity SAP) interfaces. The NLME utilizes the NLDE to achieve some of its management tasks and it also maintains a database of managed objects known as Network Information Base (NIB). In addition to these external interfaces, there is also an implicit interface between the NLME and the NLDE that allows the NLME to use the NWK data service [24].

- **Network Layer Data Entity (NLDE)**: The NLDE provides the data transmission service to allow an application to transport application protocol data units (APDU) between devices located on the same network. It provides the following services:

  - Generation of the Network level protocol data units (NPDU): The NLDE will create the NPDU from an application support sub-layer PDU by adding an appropriate protocol header.
  - Topology-specific routing: The NLDE will transmit an NPDU to another device which will be either the final destination of the NPDU or the next step towards this final destination.
  - Security: The NLDE will provide the ability to ensure the authenticity and confidentiality of a communication.

- **Network Layer Management Entity (NLME)**: The NLME provides a management service to allow the applications to interact with the stack. The services provided by this element are the following:

  - Configuring a new device: ability to configure the stack as required for operation. Among the configuration options allowed they are: beginning an operation as a ZigBee® coordinator or joining an existing network.
  - Starting a network: ability to establish a new network.
  - Joining, rejoining and leaving a network: ability to join, rejoin and leave a network, as well as the ability of a ZigBee® coordinator or router to request a device to leave the network.
  - Addressing: ability of ZigBee® coordinators and routers to assign addresses to devices joining the network.

- Neighbor discovery: ability to discover, record and report information about the one-hop neighbors of a device.
- Route discovery: ability to discover and record paths through the network by means of which messages will be routed.
- Reception control: ability for a device to control when the receiver is activated and for how long, enabling MAC sub-layer synchronization or direct reception.
- Routing: ability to use different routing mechanisms such as unicast, broadcast, multicast or many to one to efficiently exchange data in the network.

Routing Algorithms

ZigBee® routing protocol combines Cluster-Tree with AODVjr algorithms [25], which are explained in the following paragraphs.

- **AODVjr Routing algorithm**:

AODVjr is a demand-driven algorithm with some improvements over AODV (Ad hoc On-demand Distance Vector) algorithm. In fact, it reduces some features of AODV, but keeping the main function. In AODVjr only the target node can send Routing Reply Packets (RREP), avoiding circulation problems and also invalid RREP packets. Furthermore, AODVjr deletes routing error packets and precursor list. Hello packets sent by AODV periodically are also deleted in order to avoid broadcast storms.

The route discovery process is, as explained by Jianpo Li et al. [25], and shown in Fig. 1.16: node "A" wants to send a message to node "D" but it does not know the route to "D", so node "A" multicasts Routing Request (RREQ) to ask its neighbors for help to find the path to node "D". Each node receives the RREQ and keeps routing information to node "D", which will receive in this way the RREQ, and it decides whether to update its routing tables based on the RREQ routing costs after receiving it. The node "D" replies an RREP packet to node "A" through the path of minimum routing cost at the same time. Node "A" usually use multicast to send the RREQ but node "D" replies the RREP packet to "A" by unicast. The node "A" decides the best way to communicate with node "D" based on the path of minimum routing cost after receiving RREP. Now node "A" sends data to node "D" which, in turn, sends a confirmation message

- **Cluster-Tree Routing algorithm or HRT (Hierarchical Routing Algorithm)**:

The hierarchical routing algorithm, explained by Xiaohui Li et al. [26], depends on the topology and a distributed addressing scheme of ZigBee® networks, being only suitable for tree topologies.

In this algorithm, the coordinator calculates a number of parameters before the formation of the network: the maximum number of children of a router(Cm),

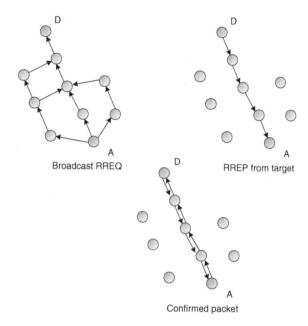

Fig. 1.16 AODVjr searching routing

the maximum number of child routers of a router(Rm), and the depth of the network(Lm). A child of a router can be a router or an end device, so $Cm \geq Rm$.

ZigBee® specifies a distributed address assignment using these parameters to calculate the nodes' network addresses. If a device joins a network successfully, it can obtain a network address from the coordinator or a router. The basic idea of the assignment is that for the coordinator and the routers in every layer, the whole address space is logically partitioned into $Rm + 1$ blocks. The first Rm blocks are to be assigned to the router child devices and the last block is reserved for the $(Cm - Rm)$ child end devices. In order to make the assignment easily, a function called $Cskip$ (formula 1) can be computed by Cm, Rm, and Lm. The value of this function is the size of the address sub-block being distributed by each parent at that depth to its router child devices for a given network depth d.

$$Cskip(d) = \begin{cases} 1 + Cm(Lm - d - 1), & \text{if } Rm = 1 \\ \frac{1 + Cm - Rm - Cm \times Rm^{Lm-d-1}}{1 - Rm}, & \text{otherwise} \end{cases} \tag{1.1}$$

$$N = \begin{cases} D, & \text{if } D > A + Rm \times Cskip(d) \\ A + 1 + floor\left(\frac{D - (A+1)}{Cskip(d)}\right), & \text{otherwise} \end{cases} \tag{1.2}$$

The address assignment algorithm is as follows (shown in Fig. 1.17):

Input: *Cm,Rm, Lm, d*
Output: *A* is the node's address

if *d = 0* **then**

$A = 0$;
else if $0 < d \leq Lm$ **then**

if the child is a router **then**

if the router is the first served **then**

$A =$ its parent address $+ 1$;
Else

$A =$ previous   router   child   device   address
$+Cskip(d-1)$;
end
else

$A = An$;
end

When ZigBee® adopted the hierarchical routing algorithm and a device called "*X*" with address "*A*" and depth "*d*" received a packet, the device extracted the destination address called "*D*". The routing procedure of the packet is described as:

Input: *A,D, d*
Output: *N* is the next hop address

if *D = A* **then**

the upper layer of *X* processes the packet;
else

if *D in interval (A,A + Cskip(d − 1)* **then**

N computed by formula 2;
Else

N=the parent's address;
end
end

ZigBee® Default routing strategy:

Applications based on ZigBee® stack often adopt the default routing strategy. It uses AODVjr by default and the hierarchical routing algorithm as last resort in the NWK layer. Generally, the nodes can be divided into two categories: nodes that have enough storage space and capacity to carry out AODVjr protocol and nodes with limited storage space which does not have this capacity.

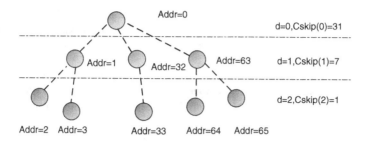

Fig. 1.17 Example of a cluster-tree address assignment for a network with Rm = 4, Cm = 6 and Lm = 3

The second kind of nodes can only use Cluster-Tree algorithm to route the packets. The packets received by this kind of nodes are sent to the next jump immediately because there is not routing discovery process and the nodes do not maintain routing tables. Therefore the routing is not always the best, being AODVjr the algorithm which calculates the optimal route.

Nefzi and Song [27] study about ZigBee® routing concluded that in the overwhelming majority of situations, the latency of AODVjr is longer than the HRT algorithm because the route discovery process of AODVjr leads to an extra delay. On the other hand, the energy consumption of the HRT is more than AODVjr because HRT is a static routing algorithm and the energy consumption in AODVjr is more uniformly distributed over the network. Considering that energy is a critical problem in WSN, the use of AODVjr by default is a justified choice despite the good performance on latency of the HRT algorithm.

- **Routing improvements in the bibliography**:

Several authors propose algorithm improvements like Xiao and Liu [21], who, in order to reduce the energy consumption of AODVjr, propose the E-AOMDVjr algorithm, an improvement of AOMDV. With this multipath energy balance routing algorithm they prove to reduce the routing overhead, avoiding the premature death of high energy nodes and making use of the network resources efficiently.

The main differences between AOMDV and AODV are: establishing and maintaining an updated mechanism for non-circulation path; and the obtaining of multiple linked-disjoint paths and a distributed protocol approach.

Finally the improvements of E-AOMDVjr with respect to AOMDV are:

1. Reducing the control overhead and simplifying the route discovery process, by no using the sequence number of destination node.
2. Simplifying the routing table structure by not having the "precursor list".
3. Using local reparation in case of interruption during data transmission. In this case only destination nodes are allowed to reply RREP because destination does not use sequence numbers.

4. Not sending HELLO packets periodically (as done in AOMDV), instead, they update neighbor node list only according to the received packet or the information from MAC layer, thus saving part of control overhead.

Furthermore, an energy balance algorithm is adopted to reduce the possibilities that some key nodes or low energy nodes to die during information transmitting, which could cause some paths to be out of date or even the whole network to stop working.

ZigBee® Application Layer

Description

The ZigBee® application layer, as explained in its specification [24], consists of several components like the APS (application support sub-layer), the ZDO (ZigBee® device objects, containing the ZDO management plane), and the manufacturer defined application objects, hosting by the Application Framework (see Fig. 1.14).4">1.14).

- **Application Support Sub-Layer**

  The application support sub-layer (APS) provides an interface between the network layer (NWK) and the application layer (APL) through a general set of services that are used by both the ZDO and the manufacturer-defined application objects. The services are provided by two entities:

  1. The APS data entity (APSDE) through the APSDE service access point (APSDE-SAP): provides the data transmission service between two or more application entities located on the same network.
  2. The APS management entity (APSME) through the APSME service access point (APSME-SAP): provides a variety of services to application objects including security services and binding of devices. It also maintains a database of managed objects, known as the APS information base (AIB).

- **Application Framework**

  The application framework in ZigBee® is the environment in which application objects are hosted on ZigBee® devices. Up to 240 distinct application objects can be defined, each identified by an endpoint address from 1 to 240. Two additional endpoints are defined for APSDE-SAP usage: endpoint 0 is reserved for the data interface to the ZDO, and endpoint 255 is reserved for the data interface function to broadcast data to all application objects. Endpoints 241–254 are reserved for future use.

  - Application profiles: Application profiles are agreements for messages, message formats, and processing actions that enable developers to create an interoperable, distributed application. These application profiles enable applications to send commands, request data, and process commands and requests.

- Clusters: Clusters are identified by a cluster identifier, which is associated with data flowing out of, or into, the device. Cluster identifiers are unique within the scope of a particular application profile.

- **ZigBee®Device Objects**:

The ZigBee® device objects (ZDO) represent a base class of functionality that provides an interface between the application objects, the device profile, and the APS. The ZDO is located between the application framework and the application support sub-layer. It satisfies common requirements of all applications operating in a ZigBee® protocol stack. The ZDO is responsible for the following:

- Initializing the application support sub-layer (APS), the network layer (NWK), and the Security Service Provider.
- Assembling configuration information from the end applications to determine and implement discovery, security management, network management, and binding management.

The ZDO presents public interfaces to the application objects in the application framework layer for control of device and network functions by the application objects. The ZDO interfaces with the lower portions of the ZigBee® protocol stack, on endpoint 0, through the APSDE-SAP for data, and through the APSME-SAP and NLME-SAP for control messages. The public interface provides address management of the device, discovery, binding, and security functions within the application framework layer of the ZigBee® protocol stack.

Device discovery is the process whereby a ZigBee® device can discover other ZigBee® devices. There are two forms of device discovery requests: IEEE address requests and NWK address requests. The IEEE address request is unicast to a particular device and assumes the NWK address is known. The NWK address request is broadcast and carries the known IEEE address as data payload.

Service discovery is the process whereby the capabilities of a given device are discovered by other devices. Service discovery can be accomplished by issuing a query for each endpoint on a given device or by using a match service feature (either broadcast or unicast). The service discovery facility defines and utilizes various descriptors to outline the capabilities of a device. Service discovery information may also be cached in the network in the case where the device proffering a particular service may be inaccessible at the time the discovery operation takes place.

Protocols

The ZigBee® application layer (ZAL), and the ZigBee® cluster library (ZCL) specify an application protocol enabling interoperability between ZigBee® devices at the application layer. The ZigBee® Alliance maintains a series of specifications for ad hoc networking between embedded devices using a single radio, IEEE

802.15.4^{TM}. Typical applications for ZigBee$^{®}$ include home automation, energy applications and similar local area wireless control applications. ZigBee$^{®}$ makes use of a vertical profile approach over the ZAL and ZCL, with profiles for different industry applications such as the ZigBee$^{®}$ home automation profile or the ZigBee$^{®}$ smart energy profile. The ZAL and ZCL provide the key application protocol functionality in ZigBee$^{®}$, enabling the exchange of commands and data, service discovery, binding and security along with profile support. These protocols use compact binary formats with the goal of fitting in small IEEE 802.15.4^{TM} frames.

6LoWPAN

6LoWPAN is a developing standard from the Internet Engineering Task Force (IETF) 6LoWPAN Working Group [28] designed from the start to be used in small/pico sensor networks.

As stated by Geoff Mulligan (from the 6LoWPAN Working Group) [29], 6LoWPAN comprises a protocol definition to enable IPv6 packets to be carried on top of low data rate, low power, small footprint radio networks (LoWPAN) as typified by the IEEE 802.15.4^{TM} radio (see Fig. 1.18). Their final goal was to define an adaptation layer to deal with the requirements imposed by IPv6, such as the increased address sizes and the 1280 byte MTU. The final design took the concepts used in IPv6 to create a set of headers that allow for the efficient encoding of large IPv6 addresses/headers into a smaller compressed header (sometimes as small as just 4 bytes), while at the same time allowing for the use of various mesh networks and supporting fragmentation and reassembly where needed.

Fig. 1.18 6LoWPAN protocol stack

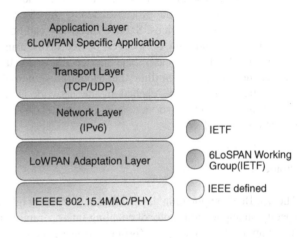

6LoWPAN Network Layer

Description

6LoWPAN incorporates a new layer between IPv6 network layer and 802.15.4™ MAC layer, called Adaptation Layer, whose main tasks are header compression, packet fragmentation, and layer two forwarding.

An IPv6 packet is too big regarding to the Maximum Transmission Unit (MTU) size of 802.15.4™ standard thus it has to be fragmented. Breaking up a big packet to number of small fragments and attaching new headers to each fragment may affect the energy level.

The Adaptation Layer must be provided to comply with the IPv6 requirements of a minimum MTU. However, it is expected that most applications of IEEE 802.15.4™ will not use such large packets, and small application payloads in conjunction with the proper header compression will produce packets that fit within a single IEEE 802.15.4™ frame.

As observed in the RFC 4944 [18], the reason of this adaptation layer is not just for IPv6 compliance, as it is quite likely that the packet sizes produced by certain application exchanges (e.g., configuration or provisioning) may require a small number of fragments.

Rather than defining a single monolithic header, as was done for IPv4 and ZigBee®, the working group [29] decided to use the same techniques that are used in IPv6 (stacked headers). Therefore, if a device is sending short packets directly to another node, it does not need to use header fields for mesh networking or fragmentation, it only requires sending the minimum encoding headers.

There are currently four basic header types defined in the standard:

- **Dispatch Header** (1 byte): define the type of header to follow
- **Mesh Header** (4 bytes) is used to standardize how to encode the hop limit and the link layer source and destination of the packet. Since 802.15.4™ allows for the use of 16 bit short addresses in addition to the standard 64 bit addresses, the Mesh Header includes two single bit fields to indicate if the originating or final address is a short or long address.
- **Fragmentation Header** (4 bytes for the first fragment and 5 bytes for subsequent fragments) supports the fragmentation and reassembly of payloads larger than the size of the 802.15.4™ frame (102 bytes of payload) and includes fields to specify the size of the original datagram as well as a sequence number for ordering the received packets. Even though for most sensor networks the application would be aware of the underlying link layer frame size limitation, in order to be compatible with the IPv6 specification the protocol was required to be able to support a minimum MTU of 1280 bytes. Therefore it was necessary to include a fragmentation and reassembly mechanism.
- **HC1 Header** (IPv6 Header Compression Header) This HC1 header allows for the stateless header compression of the IPv6 header. To accomplish this goal the protocol uses a combination of the following facts:

- the low order 64 bits of an IPv6 address (the link local address) can be the device's MAC address
- the 802.15.4™ frame carries these MAC addresses
- a number of the fields in the IPv6 header are static.

Combining all of these features allows the protocol to compress the standard 40 byte IPv6 header down to just 2 bytes.

For the simplest case the only necessary headers are a Dispatch Header, an HC1 header and the compressed IPv6 header (the total of which is 4 bytes). Only when you send large packets you need to include a Fragmentation Header, and only when you are using a mesh network layer under 6LoWPAN you need to include a Mesh Networking header. The cost of implementing 6LoWPAN is less than or at most equal to other similar protocols and the overhead for the most common packets are much less than other protocols [29]. This reduced overhead translates into energy savings.

Routing Protocols

The routing protocol in 6LoWPAN is sensitive because capabilities of nodes are limited in terms of energy, transmission range etc. Based on which layer the routing decision occurs Chowdhury et al. [22] divide routing protocols in 6LoWPAN into two categories: the mesh-under and the route-over (see Fig. 1.19).

The **mesh-under** approach performs its routing at adaptation layer and performs no IP routing within LoWPAN whereby it is directly based on the IEEE 802.15.4™ MAC addresses (16-bit or 64-bit logical address). An IP packet is fragmented by the adaptation layer to a number of fragments. These fragments are delivered to the next hop by mesh routing and eventually reach to the destination. Different fragments of an IP packet can go through different paths and they are gathered at the destination. If all fragments are reached successfully, then the adaptation layer of the destination node reassembles all fragments and creates an

Fig. 1.19 6LoWPAN routing schemes

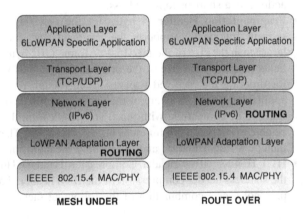

IP packet. In case of any fragment missing in the forwarding process the entire IP packet i.e. all fragments for this IP packet are retransmitted to the destination for recovery.

On the other hand, the **route-over** approach performs its routing at network layer and performs IP routing with each node serving as an IP router. The globally unique IP address of each node is created automatically by appending its interface identifier (either 16 bits or 64 bits) to the IPv6 prefix that received via router advertisement (RA). This is known as stateless auto-configuration method which is one of the 6LoWPAN features. For routing and forwarding processes the network layer takes decision using the additional encapsulated IP header. The adaptation layer of 6LoWPAN establishes a direct mapping between the frame and the IP headers. When an IP packet is fragmented by the adaptation layer, fragments are sent to the next hop based on the routing table information. The adaptation layer of the next hop checks received fragments. If all fragments are received successfully, the adaptation layer creates an IP packet from fragments and sends it to the network layer. If the packet is destined for itself, the network layer sends the IP packet to the transport layer, otherwise forwards the packet to the next hop based on the routing table information. If there are one or more fragments missing, then all fragments are retransmitted to one hop distance. After receiving all fragments successfully the adaptation layer creates an IP packet from these fragments and passes it to the network layer. The network layer then forwards or processes the IP packet based on the destination of the packet and the routing table information.

Chowdhury et al. analysis [22] over the probabilistic model of route-over and mesh-under routing schemes, concluded that **route-over** scheme is more reliable to deliver fragments from the source to the destination than **mesh**-under scheme. Furthermore, adopting selective retransmission in both route-over and mesh-under schemes resulted in route-over scheme to perform better than mesh-under scheme in terms of the total number of transmission. However, in case of total delay, mesh-under scheme performed better than route-over scheme.

Gee Keng Ee et al. [14] reviewed and compare some of the options for routing protocols in 6LoWPAN environments; the pool of routing schemes in this case is very limited due to the constrained resources of 6LowPAN devices. Some of these protocols are:

- **LOAD** (6LoWPAN Ad-Hoc On-demand Distance Vector Routing): it is a simplified on-demand routing protocol based on AODV. It is defined to be operating on top of the adaptation layer instead of the transport layer. It creates a mesh network topology underneath and without the knowledge of IPv6. IPv6 sees a 6LoWPAN as a single link. One of the differences with AODV is that it does not use destination sequence number, for ensure loop freedom, only destination generates RREP. The attributes to make a route preferable are: having the minimum number of weak links along the way, i.e., links whose LQI (Link Quality Indication) is worse than a certain threshold value, and also the minimum number of hops from source to destination. Another difference to AODV is the use of LLN (Link Layer Notification) instead of *Hello* messages.

- **DYMO-low** (Dynamic MANET On-demand for 6LoWPAN Routing): DYMO protocol is also based in AODV but unlike it, the DYMO protocol does not use local repair although it uses *Hello* messages to keep track of the link connectivity. DYMO is positioned on top of IP, using User Datagram Protocol (UDP) as the underlying protocol. However, it cannot be directly applied in 6LoWPAN routing due to its increased memory and power consumption. Thus, DYMO-low was proposed to suit DYMO into the 6LoWPAN environment. Instead of using the IP layer, DYMO-low operates on the link layer directly to create a mesh network topology of 6LoWPAN devices without the knowledge of IP, such that IP sees the WPAN as a single link. All the 6LoWPAN devices on that WPAN are on the same IPv6 link, sharing the same IPv6 prefix. DYMO-low uses 16-bit link layer short address or IEEE 64-bit extended address (EUI-64). All of the features that discussed in LOAD above are used in DYMO-low except that it uses 16-bits sequence numbers to ensure loop freedom. Besides that, LOAD's local repair and route cost accumulation are no used in DYMO-low, but it does use 16-bit destination sequence number.
- **HiLow** (Hierarchical routing): it is proposed for 6LoWPAN to increase the network scalability. Unlike AODV and LOAD that use IEEE 64-bit identifier, HiLow use 16-bit unique short address as interface identifier for memory saving and larger scalability. In HiLow, when an IEEE 802.15.4™ device (or child) wants to join a 6LoWPAN, it first tries to discover an existing 6LoWPAN by scanning procedures. If there is no 6LoWPAN in its personal operating space (POS), the child device becomes the initiator (or coordinator) of a new 6LoWPAN and assigns its short address by 0. Otherwise, the child device can find an existing neighbor device (or parent) of the existing 6LoWPAN and tries to associate with the parent at the MAC layer to receive a 16-bit short address. Every child node receives a short address by the following equation:

$$\text{Achild} \ = \ \text{Cm} \ast \text{Aparent} \ + \ \text{N} \quad (0 < \text{N£MC})$$

In the equation, *Achild* is the child node address, *Cm* is the maximum number of children a parent can have, *Aparent* is the address of the parent, and **N** is the nth child node.

For the routing operation, in HiLow, it is assumed that every node knows its own depth. When a node receives an IPv6 packet, it is called the current node. The current node determines first whether it is either the ascendant or descendant nodes of the destination by using the given equation. After that, the current node determines the next hop node to forward the packet by using the algorithm in [13]. However, when there is a link break in a route, HiLow does not support any recovery path mechanism like AODV and LOAD does.

Each routing protocol (compared Table 1.8 has its own advantages depending on the application it involves. There are some tradeoffs in the respective routing protocols such as routing protocol that uses *Hello* message may give a more reliable but higher delay in the packet routing. Although HiLow gives an advantage of memory saving to provide a larger scalability, its convergence to network

Table 1.8 6LoWPAN routing protocols comparison

	AODV (WSN)	LOAD	DYMO-low	HiLow
RERR message	Use	Use	Use	No use
Sequence number	Use	No use	Use	No use
Precursor list	Use	No use	No use	No use
Hop count	Use	Optional	Optional	Use
Hello message	Use	No use	Use	No use
Local repair	Use	Use	No use	No use
Energy Usage	High	Low	Low	Low
Memory usage	High	Medium	Medium	Low
Mobility	Mobile	Mobile	Mobile	Static
Scalability	Low	Low	Low	High
Routing delay	High	Low	High	Low
Convergence to topology change	Fast	Fast	Fast	Slow

topology change is slower compared to LOAD and DYMO-low. This will induce more delay for the route discovery process in HiLow.

6LoWPAN networking has features that place special requirements on application protocol design, although in general 6LoWPAN enables and simplifies the use of IPv6 over demanding link layers. Networking issues include the use of UDP, the compression of UDP ports and 6LoWPAN fragmentation.

According to Shelby [15], UDP has the most favorable characteristics for use over 6LoWPAN and is universally supported in protocol stacks. Although TCP has some justified use, it would require a new reliable transport or modified TCP to become universal over 6LoWPAN. Many Internet protocols today rely on TCP for a reliable connection-oriented byte stream. Instead, 6LoWPAN compatible application protocols mainly make use of UDP, which means the application protocol needs to deal with reliability if needed, out-of-order packets and datagrams rather than streams. If the UDP source or destination ports are compressed (as specified in RFC4944 [18]) then the port space can be limited down to 16 ports (ports 61616-61631). Although 6LoWPAN supports fragmentation in order to handle larger payloads coming in from outside the LoWPAN, the fragmentation of large payloads increases delay, packet loss probability and congestion.

6LoWPAN Application Layer

Description

Application layer in 6LoWPAN relays in standard internet IPv6 Network and Transport Layers, so it also uses the Internet Socket paradigm, however according to Shelby and Bormann [15] statements, application protocols used with 6LoW-PAN have special design and performance requirements.

The limitations of 6LoWPAN such as small frame sizes, limited data rates, limited memory, sleeping node cycles, along with the mobility of devices make the design of new application protocols and the adaptation of existing ones difficult. Furthermore, the autonomous nature of simple embedded devices makes auto-configuration, security and manageability the most important issues.

Application protocols used over 6LoWPAN need to take a number of requirements into account which are typically not an issue over general IP networks. These issues include:

Link layer: asymmetrical links prone to loose packages, typical payload sizes of 70–100 bytes, limited bandwidth, and no native multicast support.

Networking: the use of UDP, limited compressed UDP port space and performance issues regarding the use of fragmentation.

Host issues: unlike typical Internet hosts, 6LoWPAN hosts and networks are often mobile in nature during operation. Furthermore, battery-powered nodes use sleep periods with duty cycles often between 1–5 percent. A node may be identified in many ways, e.g., using its EUI-64, its IPv6 address or by a domain name, which should be taken into account.

Compression: header and payload compression to be used on existing protocols, and where to apply this compression, end-to-end or by an intermediate proxy.

Security: 6LoWPAN makes use of link-layer encryption which protects a single hop. Intermediate nodes are susceptible to attack, requiring sensitive application to employ end-to-end application level security. Edge routers need to implement firewalls in order to control the flow of application protocols in and out of LoWPANs.

Protocols

Shelby and Bormann also analyze in [15] protocols that are commonly used or have potential for use over 6LoWPAN.

1. **Web service protocols**

It is expected that 6LoWPAN will be integrated into the web service architecture, but the use of XML, HTTP and TCP makes the adaptation of web services challenging for LoWPAN nodes and networks.

All web services have the same basic problems for 6LoWPAN use. XML is typically too large for marking up content in the payload space available, HTTP headers have high overhead and are difficult to parse, and finally TCP has limitations of its own. There are two fundamental ways to integrate 6LoWPAN into a web service architecture: using a gateway approach or a compression approach.

- **Gateway approach**: A web service gateway is implemented at the edge of the LoWPAN, often on a local server or the edge router. The gateway makes the content and control of the devices available through a web service interface. In this approach web services actually end at the gateway. A proprietary or 6LoWPAN-specific protocol is needed between nodes and the gateway. Another disadvantage of this approach is that the gateway is dependent on the content of the application protocols, which creates scalability and evolvability problems. Each time a new use of the LoWPAN is added or the application format is modified, all gateways need to be upgraded.

- **Compression approach**: Web service format and protocols are compressed to a size suitable for use over 6LoWPAN. This can be achieved using standards, and has two forms: end-to-end and proxy. In the **end-to-end** approach the compressed format is supported by both application end-points. In the **proxy** approach an intermediate node performs transparent compression so that the Internet end-point can use standard web services.

Several technologies exist for performing XML compression. The WAP Binary XML (WBXML) format was developed for mobile phone browsers. Binary XML (BXML) from the Open Geospatial Consortium (OGC) was designed to compress large sets of geospatial data and is currently a draft proposal. General compression schemes like Fast Infoset (ISO/IEC 24824–1) work like zip for XML. Finally, the W3C is currently completing standardization of the efficient XML interchange (EXI) format, which performs compact binary encoding of XML. One suitable technology for use with 6LoWPAN is the proposed EXI standard, as it supports out-of-band schema knowledge with a sufficiently compact representation. LoW-PAN Nodes do not actually perform compression; instead they directly make use of the binary encoding for content, which keeps node complexity low.

XML compression alone only solves part of the problem. HTTP and TCP are still not suitable for use over 6LoWPAN. One commercial protocol solution called Nano Web Services (NanoWS), from Sensinode, applies XML compression in an efficient binary transfer protocol over UDP, which has been specifically designed for 6LoWPAN use [20]. The SENSEI project is also researching the use of embedded web services inside Internet based sensor networks [17]. The ideal long-term solution will be the standardization of a combination of XML binary encoding bound to a suitable UDP-based protocol.

2. MQ telemetry transport for sensor networks (MQTT-S)

The MQ telemetry transport (MQTT) is a lightweight publish/subscribe protocol designed for use in enterprise applications over low-bandwidth wide area network (WAN) links such as ISDN or GSM. The protocol was designed by IBM and is used in commercial products such as Websphere and Lotus, enjoying widespread use in M2 M applications. Although MQTT was designed to be lightweight, it requires the use of TCP, and the format is inefficient over 6LoWPAN networks.

In order to allow for MQTT to be used also in sensor networks, MQ telemetry transport for sensor networks (MQTT-S) was developed. This optimized protocol can be used over ZigBee®, UDP/6LoWPAN or any other simple network providing a bi-directional datagram service. MQTT-S is optimized for low-bandwidth wireless networks with small frame sizes and simple devices. It is still compatible with MQTT and can be seamlessly integrated with MQTT brokers using what is called an MQTT-S gateway.

3. ZigBee® compact application protocol (CAP)

The ZigBee® application protocol solution would have benefits used over standard UDP/IP communications as well, especially over 6LoWPAN, as the ZigBee® application protocol has been designed with similar requirements. A solution for using ZigBee® application protocols and profiles over UDP/IP has been proposed in [7], which is an IETF Internet draft. This specification defines how the ZAL (ZigBee® Application Layer) is mapped to standard UDP/IP primitives, enabling the use of any ZigBee® profile over 6LoWPAN or standard IP stacks. This adaptation of the ZAL for use with UDP/IP is called the compact application protocol (CAP). The function of the ZAL and ZCL (ZigBee® Cluster Library) are implemented by the CAP. The data protocol corresponds to the ZigBee® cluster library. The management protocol corresponds to the ZigBee® device profile handling binding and discovery. Finally the security protocol implements ZigBee® application sublayer (APS) security. Any ZigBee® public or private application profile can be implemented over CAP in the same way that it would use the native ZigBee® ZAL/ZCL. This allows for ZigBee® application profiles to directly be applied to IP networks.

4. Service discovery

Service discovery is used to find which services are offered, what application protocol settings they use, and at what IP address they are located. Typical protocols used for service discovery on embedded devices includes the service location protocol (SLP) described in RFC2608 [30], universal plug-n-play (UPnP) and devices profile for web services (DPWS).

For SLP to be used with 6LoWPAN, it needs optimizations because of the size of typical messages. There has been a proposal for a simple service location protocol (SSLP) [31], which provides a simple, lightweight protocol for service discovery in 6LoWPAN networks by an SSLP translation agent located on an edge router, allowing services to be discovered outside the LoWPAN network and vice versa.

5. Simple network management protocol (SNMP)

SNMP is a standard for the management of the network infrastructure and devices in IP networks. It includes an application protocol, a database schema and data objects. The current version is SNMPv3 exposes variables to a management system which can be GET or in some cases SET in order to configure or control a device. The variables exposed by SNMP are organized in hierarchies called management information bases (MIBs).

The polling approach used by SNMP is the biggest drawback of the approach. Polling approaches do not work for battery-powered LoWPAN nodes which use sleep schedules, and blindly polling for statistics (which may not have changed) creates unnecessary overhead. An event-based approach would need to be added to SNMP for applicability to 6LoWPAN management.

The suitability of using SNMPv3 with 6LoWPAN has also been analyzed in [32], which found that optimizations are needed to reduce the packet size and memory cost. Furthermore a MIB has been specified for 6LoWPAN in [33].

Jen-NET extension to 6LoWPAN

JenNet-IP is an IP-based networking solution enabling the "Internet of Things" [16]. It uses an enhanced 6LoWPAN network layer based in IEEE 802.15.4^{TM} wireless networking, with a "mesh-under" approach, offering a self-healing, self-forming, scalable and robust networking layer. It is an implementation optimized to the Jennic JN5148 wireless microcontroller.

JenNet-IP is architectured to support the requirements of low-power, low-cost wireless devices for a wide range of applications. Using open standard components such as IEEE802.15.4^{TM} MAC and PHY, 6LoWPAN, IP and UDP it enables developers to work with readily understood technologies.

The Jennic 6LoWPAN software includes [34]:

- C APIs (Application Programming Interfaces) to provide easy access to the layers of the 6LoWPAN software stack from the user application
- The required stack software to implement a 6LoWPAN system
- Jennic's proprietary JenNet wireless network protocol which is used on top of IEEE 802.15.4^{TM}

Application development tools are also provided as part of the Jennic Software Developer's Kit (SDK)

Features of Jennic 6LoWPAN software include:

- Support for wireless Star and Tree networks using JenNet (with IEEE 802.15.4^{TM})
- Socket formation and data transfer services via an IPv6 UDP socket interface
- Packet fragmentation and re-assembly (when an IP packet is too large, it must be broken up and transported in multiple IEEE 802.15.4^{TM} frames)
- IP level translation between 6LoWPAN wireless network and the Ethernet, via a wired-wireless router (Border-Router)

But it does not currently support broadcast/multicast addressing.

The Jennic 6LoWPAN software provides a complete package of the necessary components to easily and efficiently develop wireless IP network solutions.

Data Management and Storage Architectures

According to the classification made by Li et al. [35] there are several strategies for the storage of the measured data by the nodes of a wireless sensor network.

- **Centralized storage**: one of the simplest strategies. Each node transmits the collected data to a base station where it is stored for the user to request it. With this model, the inquiries for data proceeds directly to the base station with efficiency and low time-delay, as long as we can provide unrestricted energy and enough storage space to the centralized repository. But on the other hand data transmission from all the nodes to the base station means a lot of energy consumption and increases the risk of base station neighbors to become a bottleneck in a large scale network.
- **Local storage**: the data is stored in the internal memory of the nodes so the access requests must be routed to all nodes to obtain the measures. The consumption of energy in the nodes is very low, but the space limitation of the nodes causes data volatility (only the latest data is stored). Furthermore, the request broadcast to gather all the data from each node wastes a lot of energy and results in high delays and inefficiencies.
- **Distributed storage**: is a data-centric strategy. The measures data is not stored in local but in other nodes using distributed technology. Data storage and access is coordinated by an effective information mechanism, taking care of data access requirements. This strategy suits the distributed characteristics of wireless sensor network, but need intermediary information that means extra resources consumption.

Attending to the cost comparison between access and storage (computational cost could be neglected compared to communication cost), centralized storage has an access cost near to zero while the data storage cost is proportional to the production rate of data. Similarly, the storage cost of local storage is nearly zero and the access cost is proportional to the query frequency of data. In those cases where query frequency is significantly greater than production frequency, centralized storage would be the appropriated strategy while local storage would be preferable the other way round.

Distributed storage is consistent with the distributed characteristics of wireless sensor network but there are still some random and dynamic characteristics, in wireless sensor network, which limit the application scope of the distributed storage.

Regarding the query-processing methods in wireless sensor networks, Li et al. [35] identify two of them: **centralized method** and **net-in processing method**. In the centralized one, all the queries processing is completed in the base station, while on the net-in processing, it is extended from the base station to the sensor nodes. Compared to the centralized method, net-in processing method is limited by resources and, in addition, the distributed computing increases processing difficulty, making **centralized processing** the preferable option.

When designing a data storage management system, there is no need to concern about which kind of routing protocol the network uses, but the data storage strategy will be affected by the network architecture and topology (which sometimes is intimately related to the routing protocol). Li et al. [35] designed in their studies a data management system based on clustering system architecture using

Fig. 1.20 Hierarchical data storage management

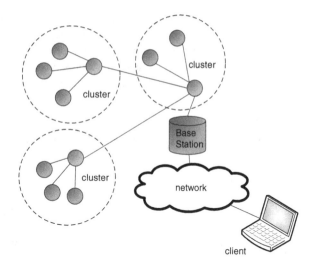

the architecture defined for clustering routing protocol. In this architecture a set of network nodes forms a cluster. Each cluster consists on a cluster head that can communicate with the base station (BS). Hence, this algorithm divides the entire network into several connected areas (see Fig. 1.20).

In this clustering management mechanism, nodes are divided into cluster head nodes and member nodes. In each cluster, the cluster head manages or control all of member nodes in the whole cluster, coordinate working between member nodes, collect responsible information inside of cluster, process data's fusion and forward data between clusters.

The **Hierarchical Data Storage Management** proposed is a combination of centralized storage, local storage and distributed storage features, based on the characteristics of clustering routing, and it is called Based-Cluster Management System for data Storage in wireless sensor network (BCDSMS).

In BCDSMS, there is a different approach to query requested data according to different data storage types. When the storage strategy is centralized, the system queries the required data from base station and passes the data to the user. For local storage of data, after receiving the data query request, the base station sends the request to the appropriate node and obtains the data that the user asked for. In distributed storage the query is sent to the data storage area to get data, or stop to spread the request when the number of hops is greater than the maximum.

BCDSMS proves a certain degree of improvement, compared to centralized and local storage, regarding energy savings and query speed.

A similar approach is used by Xiao, Luo and Chen [19], for an agricultural wireless moisture sensor network which adopts a **hierarchical storage mechanism**. In this system, leaf nodes, intermediate nodes and cluster heads store the local moisture data for the last 7 days in 512 B internal memory. Moreover, intermediate nodes and cluster heads need to save the network data packet from child nodes during the last period. Repeaters store the network data packet from cluster heads for the last 15 days in 16 KB internal programmable flash, and base station stores the moisture data in the growth period of crops such as 4 months also in flash memories. Finally data server stores all moisture data for data management, mining, and analysis and decision support.

Programming and Applications Development

Application characteristics

Embedded wireless nodes usually have a special purpose, therefore most of applications developed for wireless sensor networks will be specialized. Embedded application uses a protocol stack to configure the network and to send and receive packets. There are a number of practical issues that should be taken into account when designing an embedded application for these kinds of sensors [15].

Commissioning: In order for commissioning the link-layer used by the stack, it may be necessary for the application to configure the radio with basic parameters such as the radio channel, data rate, MAC mode and security key to enable basic connectivity with other nodes.

Device role: The node needs to configure the protocol stack for the proper type of network role, which is either a host, or a router or an edge router.

Addressing: When using the socket API of a protocol stack, the application may make use of full IPv6 addresses or just MAC addresses depending on the design. Furthermore, if the use of compressed UDP ports is desired, this requires restriction of the port space used by the application.

Mobility: The application may need to deal with some aspects of mobility, for example when the node moves from one LoWPAN to another, causing the address of the node to change.

Data reliability In case of using UDP as a transport protocol, it provides no guarantee of packet delivery order or reliability. The application must use or implement a protocol that provides sufficient reliability for its purposes.

Security: Security in 6LoWPAN networks is typically supported by the link layer as in IEEE 802.15.4^{TM}, which provides hop-by-hop encryption. However, at each router the content of the packet is vulnerable as it is decrypted at each hop. Applications that require tight security should implement end-to-end encryption of application data. Regarding ZigBee®, it provides a standardized toolbox of security specifications based on a 128-bit AES algorithm and incorporates the strong security elements of 802.15.4^{TM}. ZigBee® stack profiles define security for the MAC, network and application layers. Its security services include methods for key establishment and transport, device management and frame protection. Developers must also decide where to apply security: at the MAC, network or application layer. If the application needs the strongest security possible, secure it at the application layer. Security implemented here uses a unique session key that can only be authenticated and decrypted by another device possessing the key. This approach protects against both internal and external attacks, but it requires more memory to implement.

Power limitations: It is common in this kind of sensor networks to have a power limitation in the sensor devices. It would be interesting to consider any kind of deep low power mode in which the sensor could enter periodically. In case of solar power sensors it could be useful to consider adjustable operation modes for day and night

Development

Several platforms can be found nowadays to develop and implement wireless sensor networks (WSN) applications, each of them having its own requirements, execution and programming environments and software tools.

There are two points of view when developing WSN applications, one is the application domain expert view (oceanographers, engineers, etc.) and the second is the network experts. Application developers need to know several network specificities to build programs either by using the low-level abstractions provided by the sensor OS or directly over the hardware.

Operating Systems (OS) developed specifically for sensor nodes constitute the backbone of its architecture. Literature on embedded devices reveals the **event-based** paradigm as the prevailing model, as opposed to the **thread-based** model [1].

The **event-based** model consists of a program which is implemented as a set of independent functions or *event handlers*. Every event handler is triggered as a response to one external or internal event (e.g., hardware interruption).

In a **thread-based** system, the set of actions is performed by execution entities called *threads*, which allocate its own stack. Therefore, context switching and race conditions could happen while a thread is running, which means that inter-thread synchronization and scheduling mechanisms must be provided by the OS.

TinyOS [36] was the first open source OS specifically designed for wireless sensor devices, developed in the laboratories of the University of Berkeley by Phillip Levis under the supervision of David Culler in 2002. Soon it becomes the *de facto* standard operating system for writing sensor applications. It is also the tiniest of all existing WSN operating systems (a minimum application can occupy around 250 bytes of ROM and 16 bytes of RAM of footprint). TinyOS is written in nesC, a component-based programming language based on C that allows programming interfaces and components. Its execution model is event-based and its programming model is component-based.

Among the main disadvantages that TinyOS presents, it is worth mentioning the difficulty of maintaining or updating the application, since it is statically linked to the whole kernel.

The second version of TinyOS was released in November of 2006, and besides supporting new hardware platforms, it incorporates multiple improvements over the earlier versions and combines a thread-programming model with an event-based execution model.

Contiki [37] is a complete operating system designed for memory constrained systems, developed in Europe in the Swedish Institute of Computer Science (SICS) by Adam Dunkels as the leader of the project in 2003. It has been written in the C programming language, and allows WSN applications to be written whose typical size is around kilobytes (due to the fact that it incorporates new services). It means a bigger footprint than the applications developed in TinyOS. In spite of this, it could be considered the second most extended operating system for programming sensor nodes because it presents some contributions with respect to TinyOS:

- Applications can be more easily updated, due to the fact that Contiki supports dynamic load of programs on the top of the operating system kernel. In this way, code updates can be remotely downloaded into the network. This feature is one of the main advantages of Contiki, given that most operating systems generate an inseparable image of the system.
- Unlike most WSN operating systems, which use an event-driven programming model in order to reduce the overhead of the system, Contiki uses very light-weight threads, called *protothreads*, which can be viewed as blocking event handlers.
- It also supports a preemptive multi-thread library in the top of an event-driven kernel. This library can be included on demand by the application.

However, the most relevant contribution has been the use of *protothreads* inside Contiki.

Karpinski and Cahill [38] grouped the programming models for WSN different approaches of the literature in two categories: **macroprogramming**, where the operation of the whole network is defined as a single program, and **node-centric** programming, a more traditional approach where the focus is on the behavior of a single sensor node.

Node-centric programming allows the behavior of individual nodes to be defined, considering them as a whole. WSN operating systems have been designed to write applications on top of them, using the hardware abstractions provided. Although it is possible to obtain a certain hardware independence, operating-system independence has not been completely achieved.

Macroprogramming or network-level abstractions have the goal to realize programming from a macroscopic viewpoint that every node and data can be accessed without considering low-level communications among nodes. The term can be described as follows: "programming methodologies for sensor networks that allow the direct specification of aggregate behaviors" [1].

Wireless Sensor Networks Solutions for Smart Irrigation

In the previous part of the brief, we have elaborated in detail on various aspects of the state-of-the-art WSN solutions including major hardware solutions and main protocol stack solutions. However, the specific nature of WSN solutions required for smart irrigation calls for additional considerations. In this section, we briefly overview the major problems and solutions related to applications of WSN technology to smart irrigation systems.

Water irrigation is one of the major consumers of freshwater resources worldwide with various studies indicating as high as 70–80 % of fresh water consumption devoted only to needs of irrigation systems. Given that the worldwide reserves of fresh water are valuable resource of the mankind and are crucial for its sustainability and development, the efficient use of fresh water in irrigation systems is one of the major problems faced by water management authorities nowadays. The need for better and more efficient irrigation systems is further aggravated by the rising problem of climate changes and increased unpredictability of climate events, increased concentration of pollution due to intensive use of pesticides and herbicides, and increased agricultural production driven by needs of global population increase.

Introduction of latest technological advances is one of the most promising ways to tackle the problem of irrigation system efficiency. In this sense, WSN technology clearly represents the most promising candidate to significantly improve current irrigation systems. Soil moisture sensors are constantly improving and becoming less and less expensive and appropriate for massive deployment in WSN applications [39–40]. In combination with low-cost communication modules in sensor motes, the overall costs of WSN solution for smart irrigation application is driving the possibility for its widespread applications. However, without smart core of this system: the data analysis and decision support system, the optimal use of WSN technology is not possible. Thus the complete WSN solution for smart irrigation system represents much more than the WSN itself and strongly depends on the complex and interdisciplinary interplay of various research fields.

In general, to deploy effective WSN solution for smart irrigation system, the two main challenges exist: (1) how to deploy WSN so that it captures relevant data for smart irrigation system, and (2) how to exploit collected data within properly designed decision support system [40]. Since this brief is devoted to underlying technological enablers, we focus only on the first problem. However, it is important to note that none of these challenges can be observed independently, and the proper design of smart irrigation solution requires collaboration of different phases, from node deployment and calibration, over sensor node programming, design of data gathering phase, data collection and analysis all over to the decision support system design.

There are number of challenges identified towards the goal of deploying effective WSN solution for smart irrigation applications. These challenges lead to identification of the set of requirements WSN solution should provide, and one such exhaustive list is presented below:

- **Sensitivity**: WSN output should sensitive, i.e. susceptible to small changes in terms of soil moisture or plant demand,
- **Responsiveness**: WSN should be responsive; it should be able to provide continuous monitoring and respond rapidly (in real-time) to detected changes in order to maintain optimal water levels,
- **Universality**: WSN should be adaptable to different types of crops and different growth stages,
- **Robustness**: WSN should be robust against failures and serve as a reliable source of irrigation data,
- **Scalability**: WSN solution should be scalable and allow for initial small deployment and simple further extension towards middle-scale or large-scale deployments where necessary,
- **User-Friendly**: WSN interface towards the end-user should be intuitive, easy to use and should not require significant user training. Recent developments in smartphone applications and operative systems should serve as an excellent basis for user-friendly interface between the end-user and the WSN,
- **Actuator Powered**: WSN should represent not only sensing but acting solution as well (WSAN—Wireless Sensor and Actuator Network). WSAN is able to deliver fully automated solution, non-dependable on human labor which is expensive and potentially unreliable,
- **Energy Efficient**: Sensor nodes should be either battery-powered or preferable, should use some of the recently emerging energy harvesting solutions,
- **Reliable Communication**: WSN should provide reliable communication between sensor motes and a mote and the base station for a distances of the order of 1 km which are relevant in agricultural large-field environment,
- **Low-Cost**: WSN solution should be inexpensive and cheap to maintain and operate. Sensor nodes should be easy to deploy and replace.

Apart from WSN solution satisfying above requirements, the complete system requires appropriate use of soil humidity sensors suitable for a given application. Measuring soil moisture is typically performed either directly or via so called water balance calculations and these two approaches are the most frequent for automated irrigation applications [41].

The first approach, based on direct soil moisture measurements is the most common approach and may use various types of soil moisture sensors which measure either soil water potential (using so called tensiometers), or soil water content (using so called capacitance sensors or sensors based on time-domain reflectometry—TDR sensors). Sensors that measure water content are typically more accurate than sensors that measure water potential. Both approaches are very frequent in practice as they are easy to deploy and can be sufficiently accurate, thus justifying availability of a number of commercial solutions based on this type of sensors. However, any soil moisture sensor technology available today suffers from locality of measurements, whereas the real situation around the plant root is typically much more heterogeneous. This drives the need for sensor array that could successfully capture the variable soil moisture nature for the price of increased cost of the sensing part. Depending on the soil heterogeneity, this solution might have a significant drawback in terms of the required sensor density and thus significant deployment costs.

The second approach requires data to obtain soil moisture information indirectly, via water balance calculations. To this end, a set of sensors or measurements providing data such as rainfall and applied irrigation at one side of equation, and evaporation, run-off and drainage on the other side of equation, are required. These data can be directly measured or some appropriate available approximations could be used. However, the resulting measurements are not as accurate as direct soil moisture measurement methods. Besides the mainstream approaches described above, a number of other possibilities for soil moisture estimation are developed or currently under investigation.

WSN-based solution for smart irrigation deploys a set of sensor and actuator (such as water valve) nodes across the irrigated area. One such possible design is illustrated in Fig. 1.21 where WSN/WSAN nodes measure soil humidity and forward acquired information to the central data base over the gateway device which usually serves simultaneously the role of WSN sink as well as the gateway

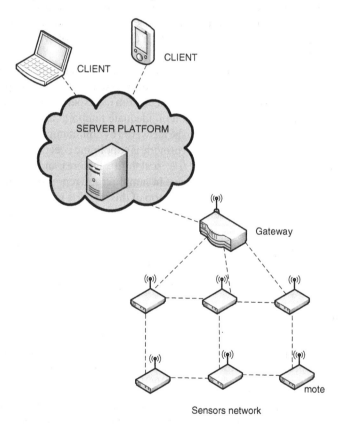

Fig. 1.21 Architecture of WSN platform for smart irrigation applications

towards the external network, most frequently mobile cellular network enabled packet services such as GPRS/EDGE. The topology of sensor network is most frequently the simplest star/tree topology, although more involved mash configurations are also possible, as depicted in the figure. Data from sensors are usually collected periodically in a push manner, i.e. the sensor node is programmed to send sensor measurements after expiration of each time period. Usually, the period of data acquisition and reporting is controllable by end user and may allow change in time-resolution of measured data. Also, data acquisition period at sensor nodes and data reporting period may differ, where typically, sensor node send averaged values of recent measurements due to the fact that communication costs significantly exceed sensing and data acquisition costs. Usual sensing period is of the order of one minute, which can be collected and reported with the same frequency or averaged to reduce the reporting frequency to periods of the order of duration of one hour. Clearly, reducing reporting period duration drastically increases battery life, but affects the time resolution of measurements which calls for carefully selected reporting period. It is highly desirable to select the operating parameters so as to extend battery life to a time span of at least one year, if not several years. Apart from soil moisture sensors, sensor nodes are typically equipped by a multitude of other sensors such as soil conductivity sensors, soil and air temperature sensors, air humidity sensors, anemometers to include effects of wind on evaporation, sunlight intensity sensors, etc.

Although the focus of this brief is on technical issues and capabilities of WSN-based smart irrigation solution, it is important to note that one of the major impacts on the successful deployment of WSN solution comes from the experts in the field of soil characteristics. In other words, the question of adequate temporal and spatial resolution of acquired data is not the question to be answered by communication engineers deploying WSN solution: what they should provide is simple and cheap possibility for deployment of additional sensor nodes (i.e. scalability feature) and easy re-programmability that would enable quick change in sampling and reporting frequencies either on all or on individual sensor nodes. On top of this, experts in soil engineering are required to provide optimal sensor densities, particular sensor locations and temporal resolution of sensor measurements and reporting. There is no need to emphasize that this task is location-specific and this kind of analysis is required to be done prior to WSN deployment, during the initial soil analysis procedure.

References

1. María Soledad Escolar Díaz "A generic software architecture for portable applications in heterogeneous wireless sensor networks", Doctoral Thesis 2010, Universidad Carlos III de Madrid. http://e-archivo.uc3m.es/bitstream/10016/9188/1/tesis-SoledadEscolar-Marzo2010.pdf (accessed March 28, 2013).
2. Costis Kompis and Simon Aliwell, Energy Harvesting Technologies to Enable Remote and Wireless Sensing, http://cmapspublic.ihmc.us/rid=1J3DN487K-21D3XZZ-1597/energyharvesting08.pdf (accessed March 28, 2013).

3. Sizing Solar Energy Harvesters for Wireless Sensor Networks, http://www.rfm.com/products/apnotes/anm1002.pdf (accessed March 28, 2013).
4. I. Stojmenovic (Ed.): "Handbook of Sensor Networks: Algorithms and Architectures," Wiley, 2005.
5. J. A. Gutierrez, E. H. Callaway, R. H. Barret, "IEEE 802.15.4™ Low-Rate Wireless Personal Area Networks: Enabling Wireless Sensor Networks," IEEE, 2003.
6. IEEE 802.15.4™-2003 Standard, 2003.
7. IEEE 802.15.4™-2006 (802.15.4™b) Standard, 2006.
8. IEEE 802.15.4™-2007 (802.15.4™a) Standard, 2007.
9. IEEE 802.15.4™c Standard, 2007.
10. IEEE 802.15.4™d Standard, 2009.
11. N. Salman, I. Rasool and A. H. Kemp, "Overview of the IEEE 802.15.4™ standards family for Low Rate Wireless Personal Area Networks," IEEE ISWCS 2010, pp. 701-705, York, UK, Sept. 2010.
12. Memsic, http://www.memsic.com/wireless-sensor-networks/ (accessed March 28, 2013).
13. K. Kim, S. Yoo, S. Daniel Park, J. Lee, G. Mulligan "Hierarchical Routing over 6LoWPAN (HiLow) draft-daniel-6lowpan-hilow-hierarchical-routing-01.txt", http://tools.ietf.org/html/draft-daniel-6lowpan-hilow-hierarchical-routing-01 (accessed March 28, 2013).
14. Gee Keng Ee, Chee Kyun Ng, Nor Kamariah Noordin and Borhanuddin Mohd. Ali"A Review of 6LoWPAN Routing Protocols" University Putra Malaysia, 2010
15. Zach Shelby, Carsten Bormann"6LoWPAN The wireless embedded internet" Ed Wiley, 2011
16. JenNet-IP wireless networking for the 'Internet of Things', http://www.jennic.com/products/protocol_stacks/jennet-ip (accessed March 28, 2013).
17. EU FP7 SENSEI Project. http://www.sensei-project.eu/ (accessed March 28, 2013).
18. RFC4944: "Transmission of IPv6 Packets over IEEE 802.15.4™ Networks",2007 https://datatracker.ietf.org/doc/rfc4944/ (accessed March 28, 2013).
19. Kehui Xiao, Xiwen Luo, Deqin Xiao, Jian Chen"Hierarchical Mechanism of Storage and Error Control for Wireless Moisture Sensor Network", 2011
20. Sensinode Oy. http://www.sensinode.com/ (accessed March 28, 2013).
21. Jun Xiao and Xiaojun Liu "The Research of E-AOMDVjr Routing Algorithm in ZigBee® Network", 2011
22. Aminul Haque Chowdhury, Muhammad Ikram, Hyon-Soo Cha, Hassen Redwan, S.M. Saif Shams, Ki-Hyung Kim, Seung-Wha Yoo "Route-over vs Mesh-under Routing in 6LoWPAN" Ajou University (Rep. of Korea) 2009
23. ZigBee® Alliance: http://www.ZigBee®.org/ (accessed March 28, 2013).
24. ZigBee® Specification, document 053474r17, 2011, http://people.ece.cornell.edu/land/courses/ece4760/FinalProjects/s2011/kjb79_ajm232/pmeter/ZigBee®%20Specification.pdf (accessed March 28, 2013).
25. Jianpo Li, Xuning Zhu, Ning Tang, Jisheng Sui. "Study on ZigBee® Network Architecture and Routing Algorithm" 2010
26. Xiaohui Li, Kangling Fang, Jinguang Gu, Liang Zhang "An Improved ZigBee® Routing Strategy for Monitoring System" 2008
27. Bilel NEFZI Ye-Qiong Song "Performance Analysis and Improvement of ZigBee® Routing Protocol" Nancy University, LORIA-INPL, 2007
28. 6LoWPAN Working Group: http://tools.ietf.org/wg/6lowpan
29. Geoff Mulligan "The 6LoWPAN Architecture" 6LoWPAN Working Group, 2007
30. RFC2608 http://www.ietf.org/rfc/rfc2608.txt (accessed March 28, 2013).
31. Simple Service Location Protocol (SSLP) for 6LoWPAN (draft), http://ebookbrowse.com/draft-daniel-6lowpan-sslp-00-pdf-d272992854 (accessed March 28, 2013).
32. SNMP optimizations for 6LoWPAN (draft), http://ebookbrowse.com/draft-hamid-6lowpan-snmp-optimizations-03-pdf-d119263899 (accessed March 28, 2013).
33. 6LoWPAN Management Information Base (draft), http://tools.ietf.org/html/draft-daniel-6lowpan-mib-01 (accessed March 28, 2013).

34. Jennic 6LoWPAN APIs User Guide 2010, http://www.jennic.com/files/support_files/JN-UG-3054-6LoWPAN-APIs-1v1.pdf (accessed March 28, 2013).
35. MA Li, TANG Changmao, WANG Chunlei, XUE Fenye "A Kind of Hierarchical Data Storage Management System Design for Wireless Sensor Network", 2011
36. TinyOs http://www.tinyos.net/ (accessed March 28, 2013).
37. Contiki http://www.contiki-os.org (accessed March 28, 2013).
38. Marcin Karpin'ski and Vinny Cahill "High-Level Application Development is Realistic for Wireless Sensor Networks", 2007
39. L. Ruiz-Garcia, L. Lunadei, P. Barreiro and J. I. Robla: "A Review of Wireless Sensor Technologies and Applications in Agriculture and Food Industry: State of the Art and Current Trends," Sensors, Vol. 9, pp. 4728-4750, 2009.
40. J. D. Lea-Cox, "Using Wireless Sensor Networks for Precision Irrigation Scheduling," in Problems, Perspectives and Challenges of Agricultural Water Management, book chapter, InTech Press.
41. Vellidis, G., V. Garrick, S. Pocknee, C. Perry, C. Kvien, M. Tucker, "How wireless will change agriculture," Stafford, J.V. (Ed.), Precision Agriculture '07 – Proceedings of the Sixth European Conference on Precision Agriculture (6ECPA), Skiathos, Greece, pp. 57-67, 2007.

Chapter 2
Sources of Remote Sensing Data for Precision Irrigation

Satellite Data

Satellite observation provides information of a large area with spatial resolution from under a meter up to 50 km and temporal resolution from 5 min up to a few days. Satellites measure radiances over various parts of the spectra of electromagnetic radiation. The data is subsequently processed to derive geophysical parameters for the observed area and delivered in this more informative form as satellite observation products. High resolution products are more suitable for irrigation [1], especially in cases where the land parcels are small. The satellite data of interest for the precision irrigation is mostly data relevant to the water cycle, hydrology and meteorology and provided by missions aimed at gathering such data.

Operational Readily Available Satellite Missions for Precision Irrigation

Currently operational satellites relevant to the problems of precision irrigation and water management include a myriad of missions run by a number of US and European agencies. We provide an overview of these missions as an integral part of the subsequent discussion of different satellite products derived from data acquired by these missions, which are directly used for precision irrigation.

The National Aeronautics and Space Administration (NASA) and United States Geological Survey (USGS) missions within the Landsat satellite series are part of a program that is the longest running enterprise for acquisition of satellite imagery of Earth. The Landsat-7 satellite with the Enhanced Thematic Mapper Plus (ETM+) was launched with the 15 April 1999 and is operational, but has experienced a malfunction that caused the loss of 22 % of the entire scene since 2003. The ETM+ is an eight-band, multispectral scanning radiometer, providing imagery with wide-ranging applications in agriculture.

D. Ćulibrk et al., *Sensing Technologies For Precision Irrigation*,
SpringerBriefs in Electrical and Computer Engineering,
DOI: 10.1007/978-1-4614-8329-8_2, © The Author(s) 2014

Another NASA mission, Earth Observing System (EOS), includes two satellites: Terra (EOS AM) and Aqua (EOS PM). Terra's orbit around the Earth is timed so that it passes from north to south across the equator in the morning, while Aqua passes south to north over the equator in the afternoon. MODIS (or Moderate Resolution Imaging Spectroradiometer) is a key instrument aboard these satellites. Terra MODIS and Aqua MODIS are viewing the entire Earth's surface every 1–2 days, acquiring data in 36 spectral bands, or groups of wavelengths. These data is aimed at improving our understanding of global dynamics and processes occurring on the land, in the oceans, and in the lower atmosphere.

In addition to MODIS, the EOS satellites have a number of other instruments of interest:

- Aqua, launched on May 4, 2002, has different instruments such as the Atmospheric Infra-red Sounder (AIRS), he Advanced Microwave Sounding Unit-A (AMSU-A) and the Cloud and the Earth's Radiant Energy System (CERES).
- Terra, launched on December 18, 1999, sports the Advanced Spaceborne Thermal Emission and Reflection Radiometer (ASTER), CERES and the Multi-angle Imaging Spectroradiometer (MISR).

AIRS and AMSU-A are aimed at deriving temperature and humidity in the lower parts of the atmosphere, while CERES is aimed at cloud monitoring, ASTER data is used to create detailed maps of land surface temperature, reflectance, and elevation. MISR is designed to measure the intensity of solar radiation reflected by the Earth system (planetary surface and atmosphere) in various directions and spectral bands.

In addition to NASA's missions, the US National Oceanic and Atmospheric Administration (NOAA) operates a series of polar-orbiting environmental satellites:

- NOAA-15 to NOAA-17 satellites (the last launched on June 24, 2002) represent a series of NOAA satellites, which carry the Advanced TIROS Operational Vertical Sounder (ATOVS), an instrument that combines High Resolution Infrared Radiation Sounder Version 3 (HIRS/3), AMSU-A and AMSU-B. In addition to ATOVS these satellites carry the Advanced Very High Resolution Radiometer (AVHRR).
- The NOAA-18 and NOAA-19, launched on May 20, 2005 and February 4, 2009 are equipped with new instruments, including the Microwave Humidity Sounder (MHS), developed by EUMETSAT, which replaced AMSU-B as part of the ATOVS on NOAA-18 and the European MetOp satellites. NOAA-19 is part of the Polar Operational Environmental Satellite (POES) system, consisting of two satellites, together with the European organization for the Exploitation of Meteorological Satellites (EUMETSAT) METOP. The system consists of a morning and afternoon satellite, in order to ensure that every part of the Earth is observed at least twice every 12 h.

Currently the ATOVS is generating products from the NOAA-15, 16, 18 and METOP-A satellites. This instrument package provides information on temperature and humidity profiles, total ozone, clouds and radiation on a global scale. The AVHRR is also a radiation-detection imager that can be used for remotely determining cloud cover and the surface temperature.

The Defense Meteorological Satellite Program (DMSP) collects and disseminates cloud cover and precipitation data acquired using a set of polar orbiting satellites. DMSP-F15, launched on December 12, 1999, carries the Special Sensor Microwave Imager (SSM/I), the Special Sensor Microwave Temperature Sounder (SSM/T-1) and the Special Sensor Microwave Water Vapour Sounder (SSM/T-2). DMSP-F15 is still operational. Subsequent missions (DMSP-F16, DMSP-F17 and DMSP-F18) are equipped with the Special Sensor Microwave Imager/Sounder (SSMIS) which replaces and improves upon the two previously used instruments. The SSMIS is able to estimate atmospheric temperature, moisture, and surface parameters from data collected at frequencies ranging from 19 to 183 GHz. All satellites are operational at the time of writing.

Suomi National Polar-orbiting Partnership, formerly known as the National Polar-orbiting Operational Environmental Satellite System (NPOESS) Preparatory Project assures a bridge between the EOS satellites and the forthcoming series of Joint Polar Satellite System (JPSS) satellites. The JPSS satellites, previously called the NPOESS are developed jointly by NASA for the NOAA. The system currently consists of a single satellite. It is equipped with several instruments relevant to irrigation. The Advanced Technology Microwave Sounder (ATMS) is a passive microwave radiometer with 22 channels dedicated to the measure of temperature and moisture profiles for weather prediction. The Cross-track Infrared Sounder (CrIS), which is a Michelson interferometer, intended for atmospheric temperature and moisture observations. The Visible Infrared Imaging Radiometer Suite (VIIRS) provide imagery of clouds under sunlit conditions in a dozen bands, and also provides coverage in a number of infrared bands for night and day cloud imaging applications. It collects visible and infrared views of Earth's dynamic and surface processes. In addition VIIRS measures atmospheric and oceanic properties, including cloud and sea-surface temperature. The Clouds and the Earth's Radiant Energy System (CERES) is a 3-channels radiometer measuring reflected solar radiation, emitted terrestrial radiation, and total radiation to monitor the Earth's total thermal radiation budget. The first satellite was launched on October 28, 2011.

The satellite missions of European organization for the Exploitation of Meteorological Satellites (EUMETSAT) of interest to the subject of this book are:

- The spin-stabilized Meteosat Second Generation (MSG) that images the full earth every 15 min and consists of a series of four geostationary meteorological satellites, which are designed to operate consecutively. This platform is equipped with the Spinning Enhanced Visible and Infrared Imager (SEVIRI) that provides imagery with a 3 km resolution at nadir and in 12 spectral bands [2]. The instrument provides data about the temperatures of clouds, land and sea surfaces, as well as allowing for the analysis of the characteristics of atmospheric air masses. The last in the series, Meteosat-10 (MSG-3) was successfully launched, on July 5th, 2012.

- The EUMETSAT Polar System (EPS) is the polar orbiting operational meteo-
rological satellite system for the morning orbit, consisting of Meteorological-
Operational (METOP) satellites, the first of which was successfully launched on
October 19, 2006. The sensors on-board METOP are the Advanced Very High
Resolution Radiometer (AVHRR), the Advanced Scatterometer (ASCAT), the
Infrared Atmospheric Sounding Interferometer (IASI), the Advanced Microwave Sounding Unit A1 and A2 (AMSU-A), the Microwave Humidity Sounder
(MHS) and the Global Navigation Satellite System Receiver for Atmospheric
Sounding (GRAS). AVHRR is a visible/infrared imaging radiometer using 6
channels for global measurement of cloud cover, sea surface temperature, ice,
snow and vegetation cover and characteristics. ASCAT is a real-aperture radar
in C-band using vertically polarized antennas for global sea surface wind vector
measurement. The main objective of IASI is to derive temperature and humidity
profiles with high vertical resolution and accuracy. HIRS is an atmospheric
sounder using 19 infrared channels (3.8–15 μm) and one visible channel to
designed to provide temperature and humidity profiles, surface temperature,
cloud parameters and total ozone. MHS is a self-calibrating, cross-track scanning, five-channel microwave, full-power radiometer operating in the
89–190 GHz range, designed to provide information on atmospheric water
vapor. The GRAS sensor uses the radio-occultation technique to retrieve temperature and humidity profiles.

The European Space Agency (ESA) main earth observation mission—the
Environmental Satellite (ENVISAT), ended on May 9, 2012. It is supposed to be
replaced by the new Sentinel satellites, to be launched during 2013. At the time of
writing the only operational ESA mission of interest is the Soil Moisture Ocean
Salinity (SMOS) Earth Explorer mission dedicated in particular to the measurement of soil moisture for hydrology applications. SMOS mission uses the
Microwave Imaging Radiometer using Aperture Synthesis (MIRAS) sensor and
was launched on November 2, 2009.

The French space agency, the Centre National d'Etudes Spatiales (CNES)
developed and runs the Satellite Pour l'Observation de la Terre (SPOT) satellite
series, which provides a very-high resolution earth observation data, appropriate
for irrigation and agriculture applications. The two operational satellites in this
mission are SPOT-4 and SPOT-5. SPOT-4 is equipped with the High Resolution
Visible and Infra-red (HRVIR) and VEGETATION sensors. SPOT-5, launched on
May 4, 2002, is equipped with High Resolution Geometric (HRG), High-Reso-
lution Stereoscopic (HRS) and VEGETATION sensors. HRVIR and HRG are 5-
band multispectral imaging sensors. The aim of the VEGETATION instrument is
to provide accurate measurements of the main characteristics of the Earth's plant
cover. Practically daily global coverage and a resolution of 1 km make this sensor
an ideal tool for observing long-term regional and global environmental changes.
HRS provides a 3D representation of the area.

Satellite Based Products for Irrigation

In this section, we provide an overview of different products that can be obtained from various readily accessible sources. The products are discussed in terms of geophysical parameters that are available within specific products, since some of them are derived from a multi-sensor approach and differ significantly in the source data obtained from the sensors.

The lowest level data acquired by the sensors is referred to as the level-1 data. The geophysical parameters are derived from the level-1 satellite data and delivered as level-2, level-3 or level-4 satellite products. A level-2 product consists of derived geophysical variables that maintain the same resolution and location as level-1 source data. The level-3 product corresponds to gridded variables in derived spatial and/or temporal resolutions. The level-4 data consists of model output or results of analyses of lower-level data.

While most of the products are distributed directly from the mission management agencies, the EUMETSAT disseminates satellite data and products both directly and through the Satellite Application Facility (SAF) network dedicated to specialized development and processing centres. Some SAFs are relevant for irrigation:

- SAF on Land Surface Analysis (LSA) develops, process and disseminates products based on MSG and MetOp data dedicated in particular to surface radiation budget and surface water balance [3].
- SAF Ocean and Sea Ice Satellite (OSI) process and disseminates products for a comprehensive information on the ocean–atmosphere interface [4].
- SAF on Support to Operational Hydrology and Water Management (H-SAF) develops process and delivers products based on microwave and infrared satellite measurements. H-SAF focuses on precipitation, soil moisture and snow cover products [5].

Surface Radiation Budget Parameters

Radiation data can be used to calculate the solar energy at ground surface, which is crucial for the estimation evapotranspiration. Such data comes in different flavors: surface temperature, Down-welling Surface Longwave Flux (DSLF), Down-welling Surface Shortwave Flux (DSSF), Downward Longwave Irradiance (DLI) and Solar Surface Irradiance (SSI).

Surface Temperature

The LSA SAF Land Surface Temperature (LST) product is estimated from MSG/SEVIRI data. It provides the LST field over Europe every 15 min. The product is operational and its resolution, determined by the SEVIRI instrument spatial resolution, is 3 × 3 km.

A surface temperature product is also generated from AVHRR/METOP data and provides LST over Europe every 12-h. The product is pre-operational and at the AVHRR spatial resolution of 1 × 1 km at nadir. The target accuracy of the product is 2 K and it is generated only in cloud free areas. In addition to temperature information, the delivered data provides quality information and uncertainty.

The surface temperature is also derived from MODIS/Aqua and Terra data. The product is operational at the spatial resolution of 5 × 5 km and provides 4 observations per day.

An LST product is be soon to be derived from VIIRS/NPP data. The spatial resolution will be increased to 750 m, with two measurements per day.

Down-Welling Surface Longwave Flux (DSLF)

The LSA SAF product is estimated from MSG/SEVIRI data and provides land surface radiation budget fields over Europe every 30 min or daily since 2005. The product is operational and at the SEVIRI spatial resolution 3 × 3 km at the sub-satellite point. A similar product is also generated by using AVHRR/METOP data, and provides DSLF over Europe every 12-h. This product is pre-operational and at the AVHRR spatial resolution of 1 × 1 km at nadir. Both products have a target accuracy of 5–10 %. They are generated only in cloud-free areas and delivered with quality information and uncertainty.

Down-Welling Surface Shortwave Flux (DSSF)

This LSA SAF product is estimated from MSG/SEVIRI data. It provides land surface radiation budget fields over Europe every 30 min or daily. The product is operational and at the SEVIRI spatial resolution of 3 × 3 km at sub-satellite point. The target accuracy is 5–10 %. It is generated only in cloud-free areas and delivered with quality information and uncertainty.

Downward Longwave Irradiance (DLI)

The OSI SAF DLI is derived from Meteosat data. The product is delivered for cloud-free areas over sea and lands in a 0.05° regular grid from 60W-60E to 60S-60N. The product is operational and is disseminated hourly and daily.

Solar Surface Irradiance (SSI)

The OSI SAF SSI is derived from Meteosat data. The product is the instantaneous field of SSI calculated over sea and land in a 0.05 degree regular grid from 60W-60E to 60S-60N. The product is operational and is disseminated hourly and daily.

The main products relevant to the surface radiative budget are listed in Table 2.1 in terms of horizontal, temporal resolution, data availability delay, condition of observation (see). Table 2.2 outlines the terms of use, shortcomings and advantages.

Surface Water Balance Parameters

Snow Cover

The snow cover information is important for the snowmelt and runoff calculation.

The snow cover LSA SAF product is estimated from MSG/SEVIRI and ASCAT/METOP data. It provides snow cover information over Europe daily (for the previous 24-h). The product is operational and distributed at the spatial resolution of 3 × 3 km at the sub-satellite point. The target accuracy is greater than 75 % hit rate in forest areas, greater than 90 % hit rate in other regions and the false alarms rate less than 3 %. The product is generated only in cloud-free areas and delivered with quality information and uncertainty.

The National Snow and Ice Data Center (NSIDC) derives snow-related information from MODIS data. The snow cover product suite is composed of products covering a range of spatial and temporal resolutions, from 500 m to 0.05 degrees, and from swath to daily, 8-day and monthly. The level 3 MODIS Aqua and Terra snow cover is a global product at the spatial resolution of 500 m gridded in a sinusoidal map projection [6, 7]. The level 2 MODIS Aqua and Terra snow cover is a 5 min swath product at the spatial resolution of 500 m [8, 9]. The product is processed daily and contains quality flags. The overall absolute accuracy of the products is about 93 %, varying by land cover and snow conditions.

Table 2.1 Main products for the surface radiative budget

Product	Horizontal resolution	Temporal resolution	Data availability delay	Condition of observation
LST SEVIRI	3 km	15 min	NRT	Cloud-free/land
LST AVHRR/MetOp	1 km	2 per day	NRT	Cloud-free/land
LST MODIS Terra/Aqua	5 km	4 per day	NRT	Cloud-free/land
DSLF SEVIRI	3 km	30 min	NRT	Cloud-free/land
DSSF SEVIRI	3 km	30 min	NRT	Cloud-free/land
DLI SEVIRI	0.05 deg	1 h	NRT	Cloud-free
SSI SEVIRI	0.05 deg	1 h	NRT	Cloud-free

Table 2.2 Licensing, advantages and shortcomings of main products for the surface radiative budget

Product	Terms of use	Shortcomings	Advantages
LST SEVIRI	License	Horizontal resolution over Europe	Temporal resolution
LST AVHRR/MetOp	License	Temporal resolution pre-operational	Horizontal resolution
LST MODIS Terra/Aqua	Free of charge	Temporal resolution	Horizontal resolution
DSLF SEVIRI	License	Horizontal resolution over Europe	Temporal resolution
DSSF SEVIRI	License	Horizontal resolution over Europe	Temporal resolution
DLI SEVIRI	License		Temporal resolution
SSI SEVIRI	License		Temporal resolution

Evapotranspiration

The LSA SAF EV product is estimated from MSG/SEVIRI data. It provides the evapotranspiration field over Europe, either every 30 min or daily. The product is in its pre-operational phase and distributed at the SEVIRI spatial resolution of 3 × 3 km at sub-satellite point. The Evapotranspiration (EV) field quantifies the water vapour flux from the ground surface (soil and canopy).The target accuracy is 25 % if evapotranspiration is greater than 0.4 mm/h and 0.1 mm/h otherwise.

The EV filed is calculated based on the reference evapotranspiration, which is the evapotranspiration from the grass reference surface as defined by the Food and Agriculture Organization (FAO). A coefficient is applied for each crop get the final EV data.

The level 4 EV is also derived from MODIS/Aqua and Terra data. The product is operational at the spatial resolution of 1 × 1 km and provides 2 observations per day in cloud-free areas. The product is disseminated with quality flags and standard deviation data.

Soil Moisture

EUMETSAT calculates the level 2 surface soil moisture index product by using ASCAT data. The product is operational and disseminated in near real time. The soil moisture index represents the degree of saturation of first layers of the soil (<5 cm) and is given in %, ranging from 0 (dry) to 100 (wet). The nominal spatial resolution is 50 × 50 km or 25 × 25 km on double swath of 550 km with a gap in between of about 700 km. A global coverage is achieved over Europe in 1.5 days. Root mean square error of 25 km resolution soil moisture index is about 0.03–0.07 m^3 of water per m^3 of the soil, depending on the soil type.

The H-SAF small-scale surface soil moisture by radar scatterometer is based on the METOP/ASCAT data. The 25 km soil moisture product is disaggregated and

Table 2.3 Main products for the surface water balance

Product	Horizontal resolution	Temporal resolution	Data availability delay	Condition of observation
Snow cover MSG+MetOp	3 km	Daily	NRT	Cloud-free
Snow cover MODIS Terra/Aqua	500 m	2 daily	NRT	Cloud-free
EV SEVIRI	3 km	30 min/daily	NRT	Cloud-free
EV MODIS Terra/Aqua	1 km	2 per day	NRT	Cloud-free
Soil moisture index ASCAT	25 km	Full coverage over Europe in 36 h	NRT	Land
Soil moisture index ASCAT+ASAR	1 km	36 h	NRT	Land

re-sampled at 1 km to satisfy hydrological requirements, using the ENVISAT ASAR data. Therefore, the spatial resolution of the product is at least 25 km and at best 1 km. The temporal resolution is 1.5 days over Europe with several gaps of coverage. The product is pre-operational, but is able to satisfy the majority of applicable requirements and is considered by the relevant steering group suitable for distribution to users, and delivered in near real time via EUMETCast and off-line via the EUMETSAT Data Centre.

The main products available for the surface water balance are listed in Table 2.3, in terms of horizontal, temporal resolution, data availability delay and conditions of observation. Table 2.4 provides the overview of the terms of use, shortcomings and advantages.

Table 2.4 Licensing, advantages and shortcomings of main products for the surface water balance

Product	Terms of use	Shortcomings	Advantages
Snow cover MSG+MetOp	License	Accuracy in forestry area/Horizontal resolution over Europe	
Snow cover MODIS Terra/Aqua	Free of charge		Horizontal resolution
EV SEVIRI	License	Horizontal resolution over Europe pre-operational	Temporal resolution
EV MODIS Terra/Aqua	Free of charge	Temporal resolution	Horizontal resolution
Soil moisture index ASCAT	License	Temporal resolution Spatial resolution	
Soil moisture index ASCAT+ASAR	License	Temporal resolution— pre-operational	Horizontal resolution

Wind-Related Products

EUMETSAT wind-related products are derived from both MSG/SEVIRI data, by tracking the motion of cloud fields and humidity patterns, and METOP ASCAT measurements. With the MSG/SEVIRI data, the height below the tropopause is determined based on infrared temperature measurements and converted into pressure levels, using the European Centre for Medium-Range Weather Forecast (ECMWF) forecasts. The Atmospheric Motion Vectors relevant to precision irrigation applications are described in the remainder of this section.

High Resolution Visible Winds (HRV)

The High Resolution Visible Winds product is calculated using a high resolution segment matrix with a 16×16 pixels segment size (about 48 km at the equator). The product is operational and generated every 1.5 h during daytime.

High Resolution Water Vapour Winds

The High Resolution Water Vapour Winds (HWW) product produces images that are divided into sub-areas of 16×16 pixels, i.e. the same resolution as the HRV product. Only the segments where a cloud has been detected are processed. The HWW is operational and generated every 1.5 h.

Ocean Surface Winds

Winds over ocean are important to quantify the ocean-atmosphere exchanges and may be of use to irrigation professionals.

The SAF OSI ASCAT 25 km wind product uses the METOP/ASCAT level 1b data at the spatial resolution of 25 km with 12.5 km cell spacing. The product gives the wind speed and direction above the ocean surface (10 m). The product is generated in near real-time, it is operational, it is disseminated every 12-h through the EUMETCast system. The product provides the wind speed in m/s from 0 to 50 m/s, but it is known to be less reliable for wind speeds in excess of 25 m/s. The product's accuracy is characterized by a wind component Root Mean Square Error (RMSE) smaller than 2 m/s and a bias of less than 0.5 m/s in wind speed.

The SAF OSI ASCAT costal wind products also use the METOP/ASCAT level 1b data at the spatial resolution of 12.5 km. The cell spacing is the same as the 25 km product but more wind data are available closer to the coast. The product gives the wind speed and direction above the ocean surface (10 m). The product is pre-operational, it is disseminated every 12-h. The product has the same characteristics as ASCAT 25 km winds product.

Table 2.5 Main wind-related products

Product	Horizontal resolution	Temporal resolution	Data availability delay	Condition of observation
HRV	48 km	1.5 h	NRT	Cloudy area daytime
HWW	48 km	1.5 h	NRT	Cloudy area
ASCAT	25 km	12 h	NRT	Ocean
ASCAT Costal	12.5 km	12 h	NRT	Ocean coast

The main wind products are listed in see Table 2.5 in terms of horizontal, temporal resolution, data availability delay, condition of observation. Table 2.6 provides the terms of use, shortcomings and advantages.

Precipitation Products

The EUMETSAT Multi-Sensor Precipitation Estimate (MPE) product is derived from MSG and SSMI/I and SSMIS instruments on DMSP satellites. The product is delivered at MSG full pixel resolution in near real-time; it consists of the rain rate in mm/h. The product is disseminated every 15 min at the GRIB2 format. The product contains two quality indicators (standard deviation and correlation coefficient) calculated on 5×5 degrees of latitude and longitude box. They are based on comparison of rain rate calculated with Meteosat 8/9 IR channels and SSM/I water vapor channels. The quality indicators are used to identify region where precipitations can be used with confidence. This product is more accurate in case of convective precipitation than in frontal or orographic rainfall. This product accuracy is good for tropical and subtropical regions and it can be used with limitations at higher latitudes.

The H-SAF precipitation rate at ground is derived form Geostationary Earth Orbiting InfraRed (IR) (MSG/SEVIRI), supported by Low Earth Orbiting microwave (DMSP/SSMIS, MetOp/MHS/AMSU-A and NOAA/MHS/AMS-A). The product is delivered at the MSG IR pixel size (average over Europe is about 8 km) over the H-SAF area limited to latitude 25–67.5 North and to longitude—25 West 45East. The accuracy is better than 10 mm/h in 80 % of the cases. This product is

Table 2.6 Licensing, advantages and shortcomings of the main wind products

Product	Terms of use	Shortcomings	Advantages
HRV	License	Horizontal resolution and sampling and only daytime	Vertical profile of wind
HWW	License	Horizontal resolution and sampling	Vertical profile of wind
ASCAT	License	Temporal resolution	
ASCAT costal	License	Temporal resolution	Winds in coastal area

not applicable for low rate (more suitable for convective precipitation). The product is pre-operational and delivered in near real-time every 15 via EUMET-Cast and EUMETSAT Data Centre.

Atmospheric Weather Parameters

Temperature, Humidity and Surface Temperature

The EUMETSAT IASI level 2 Atmospheric Water Vapour and Surface Skin Temperature (TWT) product is derived from IASI level1c, AMSU-A and MHS data. The product provides vertical profiles of temperature and humidity on 90 pressure levels, and surface skin temperature. The sampling is about 25 km at nadir. The quality of the vertical profiles is strongly related to the cloud properties available in the IASI CLP product. The accuracy of the product is 1 K for the temperature profile in the troposphere and 10 % for the relative humidity profile. The product is operational and disseminated in near real-time on GTS (The World Meteorological Organisation's Global Telecommunication System) or EUMET-cast with a timeliness of 3 h. Moreover the product is delivered through the EUMETSAT DATA Centre in HDF5 format with a timeliness of 8–9 h. the product is delivered twice per day.

The temperature and humidity profiles are derived from MODIS/Aqua and Terra data. The product provides vertical profiles on 20 pressure levels. The product is operational at the spatial resolution of 5 × 5 km and it provides 4 observations per day in cloud-free areas.

Surface Pressure

The Surface Pressure (SP) is derived from MODIS/Aqua and Terra data. The product is operational at the spatial resolution of 5 × 5 km and it provides 4 observations per day. The product is disseminated in HDF format.

The main atmospheric weather products are listed below (see Tables 2.7 and 2.8) in terms of horizontal, vertical, temporal resolution, data availability delay, condition of observation, terms of use, shortcomings and advantages.

Biogeophysical Measurements

These products are used to depict the spatial and temporal change of the vegetation cover. They are useful to initiate and update the land use variable of atmospheric models.

Table 2.7 Main atmospheric weather parameters

Product	Horizontal Resolution	Vertical resolution	Temporal resolution	Data availability delay	Condition of observation
IASI TW	12 km	90 levels	2 per day	NRT	Cloud-free
MODIS TW	5 km	12 levels	4 per day	NRT	Cloud-free
IASI skin temperature	12 km	NA	2 per day	NRT	Cloud-free
MODIS SP	5 km	NA	4 per day	NRT	Cloud-free

Table 2.8 Licensing, advantages and shortcomings of main atmospheric weather parameters

Parameter	Terms of use	Shortcomings	Advantages
IASI TW	License	Horizontal and temporal resolution	Vertical resolution
MODIS TW	Free	Temporal and vertical resolution	Horizontal resolution
IASI skin temperature	License	Horizontal and Temporal resolution	Vertical resolution
MODIS SP	Free	Temporal and vertical resolution	Horizontal resolution

Fraction of Vegetation Cover (FVC)

FVC gives the partition between soil and vegetation. This LSA SAF product is estimated from MSG/SEVIRI data. It provides vegetation cover information over Europedaily or every 10 days. The product is operational and at the SEVIRI spatial resolution 3×3 km at nadir.The product is disseminated with a target accuracy of 0.1 achieved in 75 % of land pixels. The products is generated only in cloud free and snow free areas and delivered with quality information and uncertainty. The quality of the product is bad for the large view zenith angles especially in the north of Europe.

Leaf Area Index (LAI)

The LAI is the surface of leaves per surface of ground, it measures the surface involved in radiation absorption and in vegetation atmosphere exchanges. The LAI is a key parameter for the evapotranspiration estimation. This LSA SAF product is estimated from MSG/SEVIRI data and it is processed over Europe with a characteristic time scale of 5 days and generated daily or every 10 days. The product is operational and with the spatial resolution 3×3 km at nadir. The product is disseminated with a target accuracy of 0.6 achieved in 70 % of land pixels. The products is generated only in cloud free and snow free areas and delivered with quality information and uncertainty. The quality of the product is bad for the large view zenith angles especially in the north of Europe.

The level 4 LAI is derived from MODIS/Aqua and Terra data. The product is operational at the spatial resolution of 1x1 km and it provides 2 observations per day in cloud-free areas. The characteristic time scale of the product is 8 days.

Table 2.9 Main biogeophysical parameters

Product	Horizontal resolution	Temporal resolution	Data availability delay	Condition of observation
LSA FVC	3 km	Daily	NRT	Cloud-free snow-free
LSA LAI	3 km	Daily (5 days)	NRT	Cloud-free snow-free
MODIS LAI	1 km	2 per day (8 days)	NRT	Cloud-free snow-free
LSA FPAR	3 km	Daily (5 days)	NRT	Cloud-free snow-free
MODIS FPAR	1 km	2 per day (8 days)	NRT	Cloud-free snow-free

The product is disseminated with quality flags and standard deviation. The accuracy is 0.5 LAI units RMSE globally.

Fraction of Absorbed Photosynthetic Active Radiation

Fraction of Absorbed Photosynthetic Active Radiation (FAPAR) is the PAR absorbed by the green part of the vegetation of the canopy. It is an essential parameter for the calculation of the photosynthesis and water exchange between the vegetation and the lower part of the boundary layer. This LSA SAF product is estimated from MSG/SEVIRI data and it is processed over Europe with a characteristic time scale of 5 days and generated daily or every 10 days. The product is operational and at the SEVIRI spatial resolution 3 × 3 km at nadir. The product is disseminated with a target accuracy of 0.1 achieved in 70 % of land pixels. The products is generated only in cloud free and snow free areas and delivered with quality information and uncertainty. The quality of the product is bad for the large view zenith angles especially in the north of Europe.

The level 4 FPAR is derived from MODIS/Aqua and Terra data. The product is operational at the spatial resolution of 1 × 1 km and it provides 2 observations per day in cloud-free areas. The characteristic time scale of the product is 8 days. The product is disseminated with quality flags and standard deviation. The accuracy is 0.12 FPAR units RMSE globally.

The main biogeophysical parameters are listed below (see Table 2.9) in terms of horizontal, temporal resolution, data availability delay, condition of observation, terms of use, shortcomings and advantages (Table 2.10).

Table 2.10 Licensing, advantages and shortcomings of main biogeophysical parameters

Product	Terms of use	Shortcomings	Advantages
LSA FVC	License	Horizontal resolution	
LSA LAI	License	Horizontal resolution	
MODIS LAI	Free		Horizontal resolution
LSA FPAR	License	Horizontal resolution	
MODIS FPAR	Free		Horizontal resolution

Data Accessibility

All MODIS level1 and Atmosphere data products are available to the public (at no charge) through the Level 1 and Atmosphere Archive and Distribution System (LAADS).

Selected MODIS level 4 land data are available at Land Processes Distributed Active Archive Center (LP DAAC) Data Pool. Data are downloadable via direct FTP access and at no cost to the user. The operational data are not included in this Data Pool.

The NASA's Earth Observing System Data and Information System (EOSDIS) provides access to near-real time products from the MODIS and AIRS instruments in less than 2.5 h from observation by using the Land and Atmosphere Near-real time Capability for EOS (LANCE). Data are freely available after self-registration.

The MODIS snow cover data are available from the NSIDC Data Poolweb server or by subscription for automated requests.

All data including near real-time METEOSAT, MetOp data and products delivered via EUMETCast, Direct Dissemination and FTP over the internet require a registration on the Earth Observation Portal (EO Portal). The EO Portal allows users to access and manage their subscriptions to data, products and services provided by EUMETSAT. EUMETCast is EUMETSAT's primary dissemination mechanism for the near real-time delivery of satellite data and products generated by the EUMETSAT Application Ground Segment and SAFs.

For the delivery of products, the EUMETSAT Data Policy applies. Data are subject to a set of licensing terms and conditions. The type of licence required will depend upon the *Data Usage* and the set of data you wish to receive. These conditions may involve the payment of fees. Council may waive such fees on a case by case basis for specific applications.

References

1. IEEE 802.15.4[TM]-2003 Standard, 2003.
2. IEEE 802.15.4[TM]-2006 (802.15.4[TM]b) Standard, 2006.
3. IEEE 802.15.4[TM]-2007 (802.15.4[TM]a) Standard, 2007.
4. IEEE 802.15.4[TM]c Standard, 2007.
5. IEEE 802.15.4[TM]d Standard, 2009.
6. María Soledad Escolar Díaz "A generic software architecture for portable applications in heterogeneous wireless sensor networks", Doctoral Thesis 2010, Universidad Carlos III de Madrid. http://e-archivo.uc3m.es/bitstream/10016/9188/1/tesis-SoledadEscolar-Marzo2010.pdf (Accessed March 28, 2013).
7. Sizing Solar Energy Harvesters for Wireless Sensor Networks, http://www.rfm.com/products/apnotes/anm1002.pdf (Accessed March 28, 2013).
8. Costis Kompis and Simon Aliwell, Energy Harvesting Technologies to Enable Remote and Wireless Sensing, http://cmapspublic.ihmc.us/rid=1J3DN487K-21D3XZZ-1597/energy-harvesting08.pdf (Accessed March 28, 2013).
9. Memsic, http://www.memsic.com/wireless-sensor-networks/ (Accessed March 28, 2013).

Chapter 3
GIS Applications for Irrigation Management

Although there are, as of yet, no specific GIS applications for irrigation management, there are several products related to precision agriculture which offer tools to monitor the water use in order to improve land/crop performance and offers several GIS functionalities. These precision agriculture software packages can be used to identify strategies for improved land/crop management. Precision agriculture software provides a link between crop production records and spatial (map) data, including site-specific or land management units. Precision agriculture software is required for field data collection (spatial and non-spatial), data management and data analysis.

Usually a tremendous amount of information is generated by precision agriculture, depending on the level of adoption by the farm manager or agribusiness retailer. Management changes that occur due to adoption of various precision agriculture practices must be based on accurate knowledge derived through the analysis of information. Computer software, in combination with the knowledge of farmers and specialists, will assist in determining trends and making wise decisions.

Software used in precision agriculture ranges from introductory low-end record-keeping software, to full-fledged geographic information systems (GIS), with relational database tables and full statistical analysis capabilities. Precision-agriculture software can currently be grouped into the following categories:

- farm record keeping
- yield mapping
- agricultural enhanced geographic information systems (GIS)
- geographic information systems (GIS)

People working in agribusiness can use GIS software for precision farming, land management, business operations, and much more. GIS provides the means to spatially view variables that affect crop yields, erosion and drought risk, and business opportunities. GIS has the capabilities to collect, manage, analyze, report, and share vast amounts of agricultural data to aid in discovering and establishing sustainable agriculture practices.

D. Ćulibrk et al., *Sensing Technologies For Precision Irrigation*,
SpringerBriefs in Electrical and Computer Engineering,
DOI: 10.1007/978-1-4614-8329-8_3, © The Author(s) 2014

Most available software packages provide extended GIS functionalities and store spatial data, with attribute (description) data associated with each point, line or polygon. Usually this data includes the following information:

1. Accounting data

   - debits and credits
   - balance sheet
   - net return
   - asset inventory

2. Cropping data

   - cropping history (seeding dates, harvest information)
   - field history (spraying, cultivating, water use)
   - crop inputs, (herbicides, insecticides, fertilizer)
   - weather (rainfall; temperatures, hail, snow)

3. Physical land information

   - soil type (organic matter, texture, pH, topsoil depth, parent material, classification)
   - topography (slope, aspect, curvature, position, drainage)

4. Yield-monitor data

   - volume per area
   - moisture content

5. Global positioning system (GPS) data

   - field boundary locations
   - soil sample locations
   - digital elevation information

6. Remotely sensed data

   - aerial photographs
   - satellite images

Any GIS software can usually be extended to cope with the requirements of precision irrigation solution. We provide an overview of widely used GIS solutions, both commercial and open-source.

Proprietary GIS Software

Most GIS vendors offer a complete software platform that can be used in precision agriculture. ESRI [1], one of the leading GIS vendors worldwide, offers the ArcGIS platform which includes ArcGIS Server for central geospatial data management and data distribution using web services, ArcGIS Desktop for data analysis and visualization and ArcPAD for data capture in the field. Use of this software offers many tools for spatial analysis, modeling and visualization but requires expert users, since the software is designed to serve general GIS needs. Similar GIS platforms are available from MapInfo [2] or Integraph.

ESRI ArcGIS

ESRI is a software development and services company providing Geographic Information System (GIS) software and geodatabase management applications. The headquarters of Esri is in Redlands, California. The company was founded as Environmental Systems Research Institute in 1969 as a land-use consulting firm. ESRI products (particularly ArcGIS Desktop) have one-third of the global market share. In 2002 Esri had approximately a 30 percent share of the GIS software market worldwide, more than any other vendor. Other sources estimate that about 70 percent of the current GIS users make use of ESRI products. It is the most complete and market leading solution. Therefore, we describe the system in some detail.

ESRI's **ArcGIS** is a geographic information system (GIS) for working with maps and geographic information. As all GIS it is used for: creating and using maps; compiling geographic data; analyzing mapped information; sharing and discovering geographic information; using maps and geographic information in a range of applications; and managing geographic information in a database.

The ArcGIS system provides an infrastructure for making maps and geographic information available throughout an organization, across a community, and openly on the Web.

ArcGIS includes the following Windows desktop software:

- **ArcReader**, which allows one to view and query maps created with the other ArcGIS products;
- **ArcGIS for Desktop** which supports the full functionality of ArcGIS and is the part of the suite most widely used.

ArcGIS for Desktop is licensed under three functionality levels:

- ArcGIS for Desktop Basic (formerly known as **ArcView**), which allows one to view spatial data, create layered maps, and perform basic spatial analysis;

- ArcGIS for Desktop Standard (formerly known as **ArcEditor**), which in addition to the functionality of ArcView, includes more advanced tools for manipulation of shapefiles and geodatabases; or
- ArcGIS for Desktop Advanced (formerly known as **ArcInfo**), which includes capabilities for data manipulation, editing, and analysis.

There are also server-based ArcGIS products, as well as ArcGIS products for PDAs. Such extensions can be purchased separately to increase the functionality of ArcGIS.

Desktop GIS Products

As of September 2010 Esri's current desktop GIS suite is version 10.1. ArcGIS for Desktop software products allow users to author, analyze, map, manage, share, and publish geographic information. ArcGIS for Desktop ships in three levels of licensing: ArcView, ArcEditor and ArcInfo. ArcView provides a robust set of GIS capabilities suitable for many GIS applications. ArcEditor, at added cost, expands the desktop capabilities to allow more extensive data editing and manipulation, including server geodatabase editing. ArcInfo is at the high end and provides full, advanced analysis and data management capabilities, including geostatistical and topological analysis tools. At all levels of licensing, ArcMap, ArcCatalog and ArcToolbox are the names of the applications comprising the desktop package.

ArcGIS Explorer, **ArcReader**, and **ArcExplorer** are basic freeware applications for viewing GIS data.

Many ArcGIS for Desktop Extensions are available, including Spatial Analyst which allows raster analysis and 3D Analyst which allows terrain mapping and analysis. Other more specialized extensions are available from Esri and third parties for specific GIS needs.

Esri's original product, ARC/INFO, was a command line GIS product available initially on minicomputers, then on UNIX workstations. In 1992, a GUI GIS, ArcView GIS, was introduced. Over time, both of those products were offered in Windows versions and ArcView was offered as a Macintosh product. The names ArcView and ArcInfo are now used to name different levels of licensing in ArcGIS for Desktop, and less often refer to these original software products. The Windows version of ArcGIS is now the only ArcGIS for Desktop platform that is undergoing new development for future product releases.

Server GIS Products

Server GIS products allow GIS functionality and data to be deployed from a central environment. **ArcIMS** (Internet Mapping Server) provides browser based access to GIS. **ArcSDE** (Spatial Database Engine) is used as an RDBMS connector for other Esri software to store and retrieve GIS data within a commercially available RDBMS. Currently ArcSDE can be used with Oracle, DB2, Informix, Postgresql and Microsoft SQL Server databases. It supports its native SDE binary data format, Oracle Spatial, and ST_geometry.

ArcGIS Server is the core server geographic information system (GIS) software made by Esri. ArcGIS Server is used for creating and managing GIS Web services, applications, and data. ArcGIS Server is typically deployed on-premises within the organization's service-oriented architecture (SOA) or off-premises in a cloud computing environment.

ArcGIS Server services supply mapping and GIS capabilities via ArcGIS Online for Esri Web and client applications, such as ArcGIS Desktop, ArcLogistics, the ArcGIS.com Viewer, ArcGIS Explorer, ArcGIS Explorer Online, ArcGIS Viewer for Flex, ArcGIS Mapping for SharePoint, Esri Business Analyst Online (BAO), and applications built with ArcGIS for iOS or BAO for iOS. Numerous third-party applications are licensed to use ArcGIS Server services, as well.

ArcGIS Server extensions allow GIS functionality available within a subset of ArcGIS Desktop extensions to be deployed as **Web Services**. ArcGIS Server extensions include 3D, Spatial, Geostatistical, Network, Geoportal, Image, Data Interoperability, Workflow Manager, and Schematics.

ArcGIS Server is available for the Microsoft Windows.NET Framework and the Java Platform. ArcGIS Server ships in three functional editions, Basic, Standard, and Advanced, with the Advanced edition providing the most functionality. ArcGIS Server Basic edition is used primarily to manage multiuser geodatabases and geodata services. Both ArcGIS Server Standard and Advanced editions support the following types of Web services: Feature (for Web editing), Geodata (for geodatabase replication), Geocode (for finding and displaying addresses/locations on a map), Geometry (for geometric calculations such as calculating areas and lengths), Geoprocessing (for scientific modeling and spatial data analysis), Globe (for 3D and globe rendering), Image (for serving raster data and providing control over imagery delivery, such as satellite imagery or orthophotos), Keyhole Markup Language (KML), Map (for cached and optimized map services), Mobile (for running services on field devices), Network Analyst (for routing, closest facility location, or service area analysis), Search (for enterprise search of GIS assets), Web Coverage Service (WCS), Web Feature Service (WFS) and Transactional Web Feature Service (WFS-T), and Web Map Service (WMS).

In addition, ArcGIS Server editions are available at two levels, scaled according to capacity: **Workgroup** and **Enterprise**. ArcGIS Server Workgroup can be deployed on a single machine to support a maximum of 10 simultaneous connections to a multiuser geodatabase. With Workgroup, the multiuser geodatabase storage capacity cannot exceed ten gigabytes. ArcGIS Server Enterprise supports distributed deployment of ArcGIS Server components, unlimited simultaneous connections to a multiuser geodatabase, and unlimited multiuser geodatabase storage capacity.

ArcGIS Server is also used to manage multiuser geodatabases. **Multiuser geodatabases** leverage ArcSDE technology, implemented on a relational database management system (RDBMS). ArcGIS Server Enterprise supports IBM DB2, IBM Informix Dynamic Server, Microsoft SQL Server, Oracle, and PostgreSQL. ArcGIS Server Workgroup supports Microsoft SQL Server Express R1 and R2.

ArcGIS Server is used by the software developers and Web developers to create Web, desktop, and mobile applications. Esri provides developers with application development framework (ADF) and application programming interface (API) including, ArcGIS API for JavaScript, ArcGIS API for Flex, ArcGIS API for Microsoft Silverlight/WPF, ArcGIS API for iOS, BAO API, BAO for iOS [3] as well as the ArcGIS Mobile software development kit (SDK), and ArcGIS Server REST and SOAP APIs.

Other server based products include Geoportal Extension, ArcGIS Image Server and Tracking Server, as well as several others.

Mobile GIS Products

Mobile GIS integrates GIS, GPS, location-based services, handheld computing, and the growing availability of geographic data. ArcGIS technology can be deployed on a range of mobile systems from lightweight devices to PDAs, laptops, and Tablet PCs. This suite of products: ArcPad, ArcGIS for Mobile, ArcGIS for Server (Server-oriented APIs), ArcWeb Services (Web-oriented APIs), hosted geographic databases and ArcGIS mobile.

Developer GIS Products

Developer GIS products enable building custom desktop or server GIS applications or embed GIS functionality in existing applications. These focused solutions can then be easily deployed throughout an organization. Such products include: Esri Developer Network or EDN, ArcEngine (Desktop-oriented APIs), ArcGIS for Server (Server-oriented APIs and a web development ADF which is part of ArcGIS for Server), ArcWeb Services (Web-oriented APIs)

Online GIS (ArcGIS Online)

ArcGIS includes online, Internet-based capabilities in all Esri software products. Online capabilities are centrally located at www.arcgis.com. These include web API's, hosted map and geoprocessing services, and a user sharing program. A variety of proprietary basemaps is a signature feature of arcgis.com. The Esri

Community Maps program compiles detailed user basemap information into a common cartographic format called Topographic Basemap.

Autocad Map 3D

AutoCAD® Map 3D is a mapping software from **Autodesk,** which is the market leader in the domain of Computer Aided Design (CAD) tools. The solution is intended for model-based infrastructure planning and management that helps users integrate CAD and GIS data to inform GIS, planning, and engineering decisions.

AutoCAD® Map 3D GIS software includes a rich set of survey tools to enable users to more easily import, compute, manage, and utilize field measurements acquired from GPS and terrestrial sources. The survey functionality includes the ability to consume custom field codes that can be mapped to database attributes, making asset collection more efficient using any device.

Using open-source Feature Data Object (FDO) technology, AutoCAD® Map 3D provides direct access to GIS data from a variety of spatial data sources, including ESRI® files and managed geo-databases, Oracle®, Microsoft® SQL Server®, PostGIS, PostgreSQL, SQLite, and MySQL®. It is also able to access aerial and satellite imagery and connecting to the web mapping and web feature services to take advantage of publicly available data. The open, standards-based database support enables users to join CAD objects easily to commonly used databases and spreadsheets, and to store CAD and geospatial data in relational database management and GIS systems.

Improved spatial database tools enable users to store and use multiple geometric representations of their infrastructure assets, providing with the flexibility to choose the geometry that best suits their deliverable requirements.

Analysis tools within the system help users answer questions and make decisions based on their data. The system supports: linking information in vector and tabular formats, performing data queries, creating thematic maps, building topologies, creating reports and performing buffer, tracing, and overlay analysis.

Autocad Map supports the distribution of geospatial data, maps, and designs on the web. They users can create drawings, designs, and maps, and publish them to the Internet using the Autodesk® Infrastructure Map Server application or, distribute them as individual georeferenced DWF™ files, multisheet DWF Map Books, or paper plots.

Intergraph GEOMEDIA

The **GeoMedia® product suite** is a set of integrated applications that provide users with geospatial processing capabilities. It is worth noting that the system does not rely on proprietary data, but rather accesses and uses data sources directly or data that adheres to open standards. It is an enterprise-based system, providing an organization the ability to access, conduct analysis and distribute information through the organization or over the Web.

The basic features of Geomedia platform are:

- **Universal data access**—GeoMedia's data server architecture provides access to all common geospatial forms, most computer-aided design formats, and even simple forms such as text documents.
- **Standards-based approach to enterprise and public data exchange**—The GeoMedia suite provides a strong set of interfaces for data and metadata exchange that fully align with global spatial data infrastructure standards such as those specified by **Open Geospatial Consortium (OGC)** and the **INSPIRE** Directive.
- **Easy integration with Geospatial browsers**—GeoMediaWebMap provides simple methods for integrating local data and services with the most popular geospatial browsers, such as Microsoft's Bing Maps and Google Maps.
- **Rich geospatial analysis**—The GeoMedia suite provides all the analytic and presentation tools required to enable businesses and agencies to combine their business questions with geospatial data to provide key insights for planning and efficient asset management.
- **State-of-the-art map composition**—The GeoMedia suite provides an easy-to-use, yet sophisticated map layout environment that supports workflows—from quick simple generation of workprints to complex detailed national mapping products.
- **Expansive customization environment using standard development tools**—The GeoMedia suite is designed to be extensible, using standard software development environments for the unique requirements of a particular customer workflow. The same development environment provided to customers is used to construct the products themselves, thus guaranteeing a rich and stable development platform.

GeoMedia is licensed under three functional tiers:

- GeoMedia Essentials, allows for dynamic, complex and ad-hoc query and perform basic spatial analysis of vector geospatial data across various data sources as well as create layered maps;
- GeoMedia Advantage, adds to the functionality of Essentials by adding terrain and grid functions which will allow for additional analysis on digital elevation models, flow analysis and contour line generation; or
- GeoMeida Professional, includes added capability to collect features, linear referencing management, quality assurance and validation tools as well as database schema mapping and management.

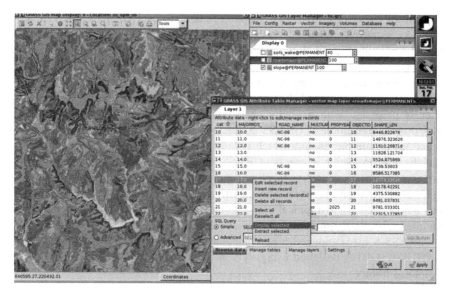

Fig. 3.1 GRASS interface screenshot

Open Source Software

While proprietary software is widely used in commercial applications, there is a number of open source GIS solutions that provide functionality needed by precision irrigation systems. We provide the overview of key aspects of the most widely used solutions.

GRASS

GRASS GIS (Geographic Resources Analysis Support System) is a free, open source desktop geographical information system (GIS) capable of handling raster, topological vector, image processing, and graphic data (Fig. 3.1).

GRASS is released under the GNU General Public License (GPL), and it can be used on multiple platforms, including Mac OS X, Microsoft Windows and Linux. Users can interface with the software features through a graphical user interface (GUI) or by "plugging into" GRASS via other software such as Quantum GIS. They can also interface with the modules directly through a bespoke shell that the application launches or by calling individual modules directly from a standard shell.

The GRASS 6 release introduced a new topological 2D/3D vector engine and support for vector network analysis. Attributes are managed in.dbf files or SQL-based DBMS such as MySQL, PostgreSQL/PostGIS, and SQLite. The system is capable of visualizing 3D vector graphics data and voxel volumes. GRASS supports an extensive range of raster and vector formats through the binding to GDAL/OGR libraries, including OGC-conformal Simple Features for interoperability with other GIS. It also supports Linear Reference System.

GRASS is designed as an environment in which tools that perform specific GIS computations are executed. Unlike GUI-based application software, the GRASS user is presented with a UNIX shell containing a modified environment that supports the execution of GRASS commands (known as modules). The environment has a state that includes such parameters as the geographic region covered and the map projection in use. All GRASS modules read this state and additionally are given specific parameters (such as input and output maps, or values to use in a computation) when executed. The majority of GRASS modules and capabilities can be operated via a graphical user interface (provided by a GRASS module), as an alternative to manipulating geographic data in a shell.

GRASS 6.4.0 introduced a new generation of graphical user interface called wxGUI. wxGUI is designed using Python programming language and wxPython graphical library.

Quantum GIS

Quantum GIS (QGIS) is a cross-platform free software desktop geographic information systems (GIS) application that provides data viewing, editing, and analysis capabilities.

Quantum GIS is written in C++, and uses the Qt library extensively. Quantum GIS allows integration of plugins developed using either C++ or Python. In addition to Qt, required dependencies of Quantum GIS include GEOS and SQLite. GDAL, GRASS GIS, PostGIS, and PostgreSQL are also recommended, as they provide access to additional data formats (Fig. 3.2).

Quantum GIS runs on different operating systems including Mac OS X, Linux, UNIX, and Microsoft Windows. For Mac users, the advantage of Quantum GIS over GRASS GIS is that it does not require the X11 windowing system in order to run, and the interface is much cleaner and faster. Quantum GIS can also be used as a graphical user interface to GRASS. Quantum GIS has a small file size compared to commercial GIS's and requires less RAM and processing power; hence it can be used on older hardware or running simultaneously with other applications where CPU power may be limited.

Quantum GIS is maintained by an active group of volunteer developers who regularly release updates and bug fixes. Currently, developers have translated Quantum GIS into 31 languages and the application is used internationally in academic and professional environments.

Quantum GIS allows use of shapefiles, coverages, and personal geodatabases. MapInfo, PostGIS, and a number of other formats are supported in Quantum GIS [4].

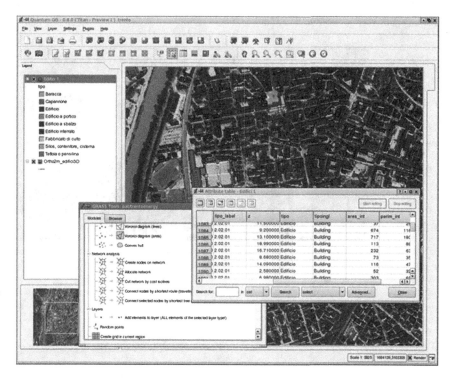

Fig. 3.2 Quantum GIS interface screenshot

Web services, including Web Map Service and Web Feature Service, are also supported to allow use of data from external sources.

Quantum GIS provides integration with other open source GIS packages, including PostGIS, GRASS, and MapServer to give users extensive functionality. Plugins, written in Python, extend the capabilities of QGIS. There are plugins to geocode using the Google Geocoding API, perform geoprocessing (fTools) similar to the standard tools found in ArcGIS, interface with PostgreSQL and MySQL databases, and use Mapnik as a map renderer.

uDIG

uDig is a GIS software program produced by a community led by Canadian-based consulting company Refractions Research. It is based on the Eclipse platform and features full layered Open Source GIS. It is written in Java and released under GNU Lesser General Public License (Fig. 3.3).

uDig can use GRASS for complex vector operations and also embeds JGRASS and specialized hydrology tools from the Horton Machine. It supports shapefiles, PostGIS, WMS, and many other data sources natively.

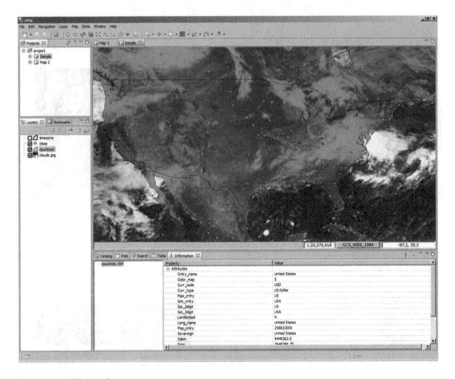

Fig. 3.3 uDIG interface screenshot

gvSIG

gvSIG is a geographic information system that is a desktop application designed for capturing, storing, handling, analyzing and deploying any kind of referenced geographic information in order to solve complex management and planning problems. gvSIG is has a simple, user-friendly interface and is able to access the most common formats, both vector and raster ones. It features a wide range of tools for working with geographic-like information. gvSig is open source software, distributed under the GNU General Public License (GPL).

It is developed using Java and is available for Linux, Windows and Mac OS X platforms (Fig. 3.4).

Outstanding gvSIG features are:

- Integrating in the same view both local (files, databases) and remote data through OGC standards.
- Being designed to be easily extendable, allowing continuous application enhancement, as well as enabling the development of tailor-made solutions.

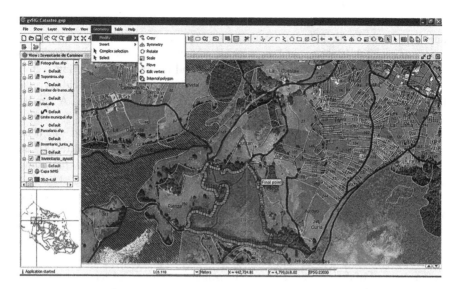

Fig. 3.4 gvSIG interface screenshot

- Being available in several languages: Spanish, English UK, English USA, German, French, Italian, Portuguese, Portuguese-Brazilian, Russian, Chinese, Serbian, Swahili, Turkish, Czech, Polish, Romanian, Greek, Basque, Valencian, Gallego.

GIS Software for Precision Agriculture

Several precision agriculture software solutions are designed to extend the basic GIS packages such as ArcView or MapInfo, in order to support precision agriculture. For example, SSToolbox [5] is built on ArcView engine.

GIS software usually supports the development of specific modules or plug-ins that can be used to increase the functionality of the basic system. Overall, software packages are becoming easier to customize to the user's needs. With more producers and agribusiness adopting components of precision agriculture there is more user feedback to the software developers as to what components and features are necessary.

In order to facilitate the use of the GIS solutions by non-expert users, several companies offer customized software solutions intended for precision agriculture. These solutions are usually standalone, modular desktop applications which focus on the visualization of the agricultural related data as well as on providing spatial analysis capabilities like surface or contours creation. The users, depending on their needs, use the related software modules (e.g. field records, accounting, mapping, analysis etc.).

In order to acquire data from sensors they provide import utilities so there is no real time connection to the field information. Recently some systems started providing access to external databases such as weather prediction or soil databases. Collaboration with mobile devices for in-field mapping and information collection is also available in some platforms. Farm Works office solutions [6] include field records, farm accounting, herd management, mapping, and water management. A similar solution is available from Agvance [7].

Recently a few solutions have come about that are Internet-based. Such an approach is followed by SST Software. SST Software offers FarmRite [8], an integrated platform that can facilitate the precision agriculture needs by providing access to several web based services. The company offers to the users, desktop and mobile interfaces in order to collect data, a cloud-based platform to store, visualize and analyze data, as well as expert services in order to provide more complex products on demand. These are provided as a result of advanced spatial analysis.

An approach similar to that of the SST has been adopted by AgriCon [9]. The AgriCon precision farming data portal offers an online platform for farmers to save and manage all of their data in a portal independent of individual service providers, software providers and manufacturers of agricultural technology.

SSToolBox

Developed by the SST Software, SSToolbox is a GIS for crop consultants, fertilizer/chemical dealers, educators, researchers, and farm managers. SSToolbox uses the farm field boundary to organize all pertinent information relating to the field, such as soil types, soil fertility, yield results, hybrid/variety selections, remotely-sensed imagery, aerial photos, chemical applications, field scouting, and more.

The solution provides analysis for decision support, yield data processing tools and creates fertility recommendations.

SSToolKit

Also provided by SST Software, SSToolkit is an economical GIS designed for farmers who have access to the more comprehensive analytical services provided by their local input supplier or crop consultant. Similar to the SSToolBox, the SSToolkit uses the farm field boundary to organize all pertinent information relating to the field and supports functionality similar to that of the SSToolBox.

Farm Works Office Desktop

Farm Works Software company, provides a suite of tools for farmers:

- Farm Works™ View software is a tool for reading and writing job data to popular precision farming devices.
- Farm Works™ Mapping software is a mapping and field record keeping solution.
- Farm Works™ Accounting software provides a general ledger and enterprising solution while maintaining a complete field record keeping system.
- Farm Works™ Surface software is an analysis and design tool for use with the Trimble® WM-Drain™ farm drainage solution. Surface ensures the optimal placement of tile and surface drains in both surface and sub-surface drainage water management projects, helping to drain fields adequately and increase crop yields.

The system provides support for keeping track of information relevant to: planting, application, harvest, soil analysis, field records, satellite imagery, etc.

Agvance Mapping

Provided by SSI, Agvance Mapping is a GIS solution intended for analysis and site-specific mapping, aimed at agricultural retailers. It can generate industry files to facilitate precision agriculture applications while integrating the use of maps throughout the entire farming operation.

The functionality provided by the Agvance Mapping includes:

- A mapping interface for the creation of boundaries and related map layers.
- Interactive tree navigation to facilitate switching between farm, grower and field level data.
- Customizable surface maps and thematic layers that can be automatically generated from spatial data stored in Agvance or dynamically from imported data.
- Soil types and related label attributes can be imported to create soil layer polygons within the current selected field or farm.
- Harvest data can be imported for historical analysis and to influence future recommendations
- Soil tests and recommendations can easily be imported from labs, consultants, and/or in industry supported formats.
- Flexible grower map books can be generated to include soil types, test levels, surface maps, yield maps and as-applied data with several printing options.

Achieving Interoperability in Precision Irrigation Support Systems

While there are a number of ways in which a state-of-the art irrigation management solution could exchange data with other GIS systems, most such systems (and all major ones) support the Open Geospatial Consortium (OGC) standard. We discuss this technology in more detail, as we feel that in order to achieve best interoperability any irrigation management solution should support this standard.

The OGC, an international voluntary consensus standards organization, originated in 1994. In the OGC, more than 400 commercial, governmental, nonprofit and research organizations worldwide collaborate in a consensus process encouraging development and implementation of open standards for geospatial content and services, GIS data processing and data sharing.

Standards

Most of the OGC standards depend on a generalized architecture captured in a set of documents collectively called the *Abstract Specification*, which describes a basic data model for representing geographic features. Atop the Abstract Specification members have developed and continue to develop a growing number of specifications, or standards to serve specific needs for interoperable location and geospatial technology, including GIS.

More information here about the standards can be accessed at [10].

The OGC standards baseline is comprised of more than 30 standards, including:

- CSW—Catalog Service for the Web: access to catalog information
- GML—Geography Markup Language: XML-format for geographical information
- GeoXACML—Geospatial eXtensible Access Control Markup Language (as of 2009 in the process of standardization)
- KML—Keyhole Markup Language: XML-based language schema for expressing geographic annotation and visualization on existing (or future) Web-based, two-dimensional maps and three-dimensional Earth browsers
- Observations and Measurements
- OGC Reference Model—a complete set of reference models
- OWS—OGC Web Service Common
- SOS—Sensor Observation Service
- SPS—Sensor Planning Service
- SensorML—Sensor Model Language
- SFS—Simple Features—SQL
- Styled Layer Descriptor

- WCS—Web Coverage Service: provides coverage objects from a specified region
- WFS—Web Feature Service: for retrieving or altering feature descriptions
- WMS—Web Map Service: provides map images
- WMTS—Web Map Tile Service: provides map image tiles
- WPS—Web Processing Service: remote processing service
These standards were originally built on the **HTTP web services** paradigm for message-based interactions in web-based systems. However, in the last year the members have started working on defining a common approach for **SOAP** protocol and **WSDL** bindings. Considerable progress has been made in defining Representational State Transfer (**REST**) web services.

The OpenGIS® Web Map Service Interface Standard

The OpenGIS® Web Map Service Interface Standard (**WMS**) provides a simple HTTP interface for requesting geo-registered map images from one or more distributed geospatial databases. A WMS request defines the geographic layer(s) and area of interest to be processed. The response to the request is one or more geo-registered map images (returned as JPEG, PNG, etc.) that can be displayed in a browser application. The interface also supports the ability to specify whether the returned images should be transparent so that layers from multiple servers can be combined or not.

The OpenGIS® Web Feature Service Interface Standard

The OpenGIS® Web Feature Service Interface Standard (**WFS**) provides an interface allowing requests for geographical features across the web using platform-independent calls. One can think of geographical features as the "source code" behind a map, whereas the WMS interface or online mapping portals like Google Maps return only an image, which end-users cannot edit or spatially analyze. The XML-based GML furnishes the default payload-encoding for transporting the geographic features, but other formats like shapefiles can also serve for transport. In early 2006, the OGC members approved the OpenGIS GML Simple Features Profile. This profile is designed to both increase interoperability between WFS servers and to improve the ease of implementation of the WFS standard.

The WFS specification defines interfaces for describing data manipulation operations of geographic features. Data manipulation operations include the ability to:

- Get or query features based on spatial and non-spatial constraints.
- Create a new feature instance.
- Delete a feature instance.
- Update a feature instance.

The OpenGIS® Web Coverage Service Interface Standard

The OpenGIS® Web Coverage Service Interface Standard (WCS) provides an interface allowing requests for geographical coverages across the web using platform-independent calls. The coverages are objects (or images) in a geographical area. The basic Web Coverage Service allows querying and retrieval of coverages.

A WCS describes discovery, query, or data transformation operations. The client generates the request and posts it to a web feature server using HTTP. The web feature server then executes the request. The WCS specification uses HTTP as the distributed computing platform, although this is not a hard requirement.

There are two encodings defined for WCS operations:

- XML (amenable to HTTP POST/SOAP).
- Keyword-Value pairs (amenable to HTTP GET/Remote procedure call).

Internet Services Interoperability

The OpenGIS® Web Processing Service Interface Standard

The OpenGIS® Web Processing Service (WPS) Interface Standard provides rules for standardizing how inputs and outputs (requests and responses) for invoking geospatial processing services, such as polygon overlay, as a Web service. The WPS standard defines how a client can request the execution of a process, and how the output from the process is handled. It defines an interface that facilitates the publishing of geospatial processes and clients' discovery of and binding to those processes. The data required by the WPS can be delivered across a network or they can be available at the server. WPS can describe any calculation (i.e. process) including all of its inputs and outputs, and trigger its execution as a Web service. WPS supports simultaneous exposure of processes via HTTP GET, HTTP POST, and SOAP, thus allowing the client to choose the most appropriate interface mechanism. The specific processes served up by a WPS implementation are defined by the owner of that implementation. Although WPS was designed to work with spatially referenced data, it can be used with any kind of data.

WPS makes it possible to publish, find, and bind to processes in a standardized and thus interoperable fashion. Theoretically, it is transport/platform neutral (like SOAP), but in practice it has only been specified for HTTP.

The OpenGIS® Sensor Observation Service

The OpenGIS® Sensor Observation Service (SOS) standard is applicable to use cases in which sensor data needs to be managed in an interoperable way. This standard defines a Web service interface which allows querying observations, sensor metadata, as well as representations of observed features. Further, this standard defines means to register new sensors and to remove existing ones. Also, it defines operations to insert new sensor observations. This standard defines this functionality in a binding independent way; two bindings are specified in this document: a KVP binding and a SOAP binding.

Observations and Measurements: XML Implementation

This standard specifies an XML implementation for the OGC and ISO Observations and Measurements (O&M) conceptual model (OGC Observations and Measurements v2.0 also published as ISO/DIS 19156), including a schema for Sampling Features. This encoding is an essential dependency for the OGC Sensor Observation Service (SOS) Interface Standard. More specifically, this standard defines XML schemas for observations, and for features involved in sampling when making observations. These provide document models for the exchange of information describing observation acts and their results, both within and between different scientific and technical communities.

OpenGIS® Sensor Model Language (SensorML)

The OpenGIS® Sensor Model Language Encoding Standard (SensorML) specifies models and XML encoding that provide a framework within which the geometric, dynamic, and observational characteristics of sensors and sensor systems can be defined. There are many different sensor types, from simple visual thermometers to complex electron microscopes and earth observing satellites. These can all be supported through the definition of atomic process models and process chains. Within SensorML, all processes and components are encoded as application schema of the Feature model in the Geographic Markup Language (GML) Version 3.1.1. This is one of the OGC Sensor Web Enablement (SWE) [11] suite of standards. For additional information on SensorML [12].

OpenGIS® Sensor Planning Service Implementation Standard

The OpenGIS® Sensor Planning Service Interface Standard (SPS) defines interfaces for queries that provide information about the capabilities of a sensor and how to task the sensor. The standard is designed to support queries that have the following purposes: to determine the feasibility of a sensor planning request; to submit and reserve/commit such a request; to inquire about the status of such a request; to update or cancel such a request; and to request information about other OGC Web services that provide access to the data collected by the requested task. This is one of the OGC Sensor Web Enablement (SWE) [11] suite of standards.

OpenGIS® SWE Service Model Implementation Standard

This standard currently defines eight packages with data types for common use across OGC Sensor Web Enablement (SWE) services. Five of these packages define operation request and response types. The packages are:

1) **Contents**—Defines data types that can be used in specific services that provide (access to) sensors;
2) **Notification**—Defines the data types that support provision of metadata about the notification capabilities of a service as well as the definition and encoding of SWES events;
3) **Common**—Defines data types common to other packages;
4) **Common Codes**—Defines commonly used lists of codes with special semantics;
5) **DescribeSensor**—Defines the request and response types of an operation used to retrieve metadata about a given sensor;
6) **UpdateSensorDescription**—Defines the request and response types of an operation used to modify the description of a given sensor;
7) **InsertSensor**—Defines the request and response types of an operation used to insert a new sensor instance at a service;
8) **DeleteSensor**—Defines the request and response types of an operation used to remove a sensor from a service.

These packages use data types specified in other standards.

OpenGIS® Web Coverage Processing Service (WCPS) Language Interface Standard

The OpenGIS® Web Coverage Service Interface Standard (WCS) defines a protocol-independent language for the extraction, processing, and analysis of multi-dimensional gridded coverages (see: http://www.opengeospatial.org/ogc/glossary/c) representing sensor, image, or statistics data. Services implementing this language provide access to original or derived sets of geospatial coverage information, in forms that are useful for client-side rendering, input into scientific models, and other client applications. Further information about WPCS can be found at the WCPS Service page of the OGC Network [13].

Impact of Latest European Directives and Projects

The precision irrigation and agriculture systems are intimately connected and should be synced with several recent global and EU directives, initiatives and projects. These initiatives deal mostly with the agriculture and water management as an integral part of environment and Earth observation topic. In particular, there is a global push to make earth observation systems interoperable. We discuss briefly the most prominent of these initiatives and projects.

INSPIRE

The EU INSPIRE Directive [14], which entered into force in May 2007, establishes an infrastructure for spatial information in Europe to support Community environmental policies, and policies or activities which may have an impact on the environment. Its goal is to ensure cross-border interoperability.

INSPIRE is based on the infrastructures for spatial information established and operated by the 27 Member States of the European Union. The Directive addresses 34 spatial data themes needed for environmental applications, with key components specified through technical implementing rules. This makes INSPIRE a unique example of a legislative "regional" approach.

Legislation

Directive 2007/2/EC of the European Parliament and of the Council of 14 March 2007 establishing an Infrastructure for Spatial Information in the European Community (INSPIRE) was published in the official Journal on the 25th April 2007. The INSPIRE Directive entered into force on the 15th May 2007.

To ensure that the spatial data infrastructures of the Member States are compatible and usable in a Community and transboundary context, the Directive requires that common Implementing Rules (IR) are adopted in a number of specific areas (Metadata, Data Specifications, Network Services, Data and Service Sharing and Monitoring and Reporting). These IRs are adopted as Commission Decisions or Regulations, and are binding in their entirety. The Commission is assisted in the process of adopting such rules by a regulatory committee composed of representatives of the Member States and chaired by a representative of the Commission (this is known as the Comitology procedure).

Legal acta relevant to INSPIRE (and, therefore, precision irrigation) include:

- Directive 2007/2/EC of the European Parliament and of the Council of 14 March 2007 establishing an Infrastructure for Spatial Information in the European Community (INSPIRE) 14.03.2007
- INSPIRE Metadata Regulation 03.12.2008
- Commission Decision regarding INSPIRE monitoring and reporting 05.06.2009
- Commission Regulation (EC) No 976/2009 of 19 October 2009 implementing Directive 2007/2/EC of the European Parliament and of the Council as regards the Network Services 19.10.2009
- Corrigendum to INSPIRE Metadata Regulation 15.12.2009
- Regulation on INSPIRE Data and Service Sharing 29.03.2010
- Commission Regulation amending Regulation (EC) No 976/2009 as regards download services and transformation service 10.12.2010
- COMMISSION REGULATION implementing Directive 2007/2/EC of the European Parliament and of the Council as regards interoperability of spatial data sets and services 10.12.2010
- COMMISSION REGULATION amending Regulation 1089/2010 as regards interoperability of spatial data sets and services 05.02.2011

GEOSS

The Global Earth Observation System of Systems (GEOSS) is in many ways a global initiative equivalent/complementary to INSPIRE, where the intention is to build a global data-sharing system. This is an initiative by the Group on Earth Observations (GEO) and the system is being developed on the basis of a 10-Year Implementation Plan running from 2005 to 2015.

GEOSS seeks to connect the producers of environmental data and decision-support tools with the end users of these products, with the aim of enhancing the relevance of Earth observations to global issues. The result is to be a global public infrastructure that generates comprehensive, near-real-time environmental data, information and analyses for a wide range of users.

Agriculture is stated as the first of GEOSS societal benefits. The GEOSS is being constructed to help farmers, fishers and policymakers maximize productivity and food security while preserving ecosystems and biodiversity. GEO also aims to support the sustainable management of agriculture by disseminating weather forecasts, early warnings of storms and other extreme events, water pollution, long-term forecasts of likely climate change impacts, and information on water supplies. These and other data are being integrated so that they can be used in models for simulating and predicting agricultural trends. Related activities include mapping the changing distribution of croplands around the world, advancing the accuracy of measurements of biomass (the total amount of living material in a given habitat or population), reporting agricultural statistics in a more timely manner, and improving forecasts of shortfalls in crop production and food supplies.

GIGAS

GIGAS (GEOSS, INSPIRE and Global Monitoring for Environment and Security (GMES) an Action in Support) was a European project co-funded by the European Commission as a Support Action under Grant Agreement number 224274 in the period June 2008 to May 2010 [15].

The initiative of GIGAS continues to promote the coherent and interoperable development of the GMES, INSPIRE and GEOSS initiatives through their concerted adoption of standards, protocols, and open architectures. Given the complexity and dynamics of each initiative and the large number of stakeholders involved, the key added value of GIGAS has been on bringing together the leading organizations in Europe who are able to make a difference and achieve a truly synergistic convergence of the initiatives. Among them, the Joint Research Centre is the technical coordinator of INSPIRE, the European Space Agency is responsible for the GMES space component, and both organizations together with a third partner, the Open Geospatial Consortium play a leading role in the development of the GEOSS architecture and components. This core group is supported by key industrial players in the space and geographic information sectors, with the scientific leadership of the Fraunhofer Institute.

This consortium followed the project objectives set through an iterative and consensus-based approach which includes: in-depth analysis of the requirements and barriers to interoperability in each of the three initiatives and strategic FP 6/FP 7 projects; comparative evaluation of this activity as input to a forum of key stakeholders at a European level; consensus building in the forum on how to update and integrate the architectures of GMES, INSPIRE and GEOSS, and influence standards development and adoption. From these recommendations followed actions to shape the direction of the initiatives and to define a roadmap for future development, including the key research topics to be addressed to sustain the convergence of the initiatives. GIGAS results and recommendations contributed to the emergence of a collaborative information space for accessing and sharing distributed environmental resources in Europe. This represents a milestone towards building a Single Information Space in Europe for the Environment.

EuroGEOSS

EuroGEOSS is a large scale integrated project in the Seventh Framework programme of the European Commission [4]. It is part of the thematic area: "ENV.2008.4.1.1.1: European environment Earth observation system supporting INSPIRE and compatible with GEOSS". EuroGEOSS demonstrates the added value to the scientific community and society of making existing systems and applications interoperable and used within the GEOSS and INSPIRE frameworks.

e project builds an initial operating capacity for a European Environment Earth Observation System in the three strategic areas of Drought, Forestry and Biodiversity. It then undertakes the research necessary to develop this further into an advanced operating capacity that provides access not just to data but also to analytical models made understandable and useable by scientists from different disciplinary domains.

This concept of inter-disciplinary interoperability requires research in advanced modelling from multi-scale heterogeneous data sources, expressing models as workflows of geo-processing components reusable by other communities, and ability to use natural language to interface with the models. The extension of INSPIRE and GEOSS components with concepts emerging in the Web 2.0 communities in respect to user interactions and resource discovery, also supports the wider engagement of the scientific community with GEOSS as a powerful means to improve the scientific understanding of the complex mechanisms driving the changes that affect our planet. Ultimately, EuroGEOSS outcomes will be extended to issues on a global scale through our sustainability initiative to support user understanding and applications for improving societal conditions.

Of specific interest to precision irrigation is the European Drought Observatory that EuroGEOSS will develop within the framework of INSPIRE specifications and GEOSS interoperability arrangements. The project will establish a drought metadata catalog to allow access to resources for drought monitoring. It will be fully integrated with local and national systems in Europe and international drought early warning systems as a European contribution to a Global Drought Early Warning System.

References

1. Pitney-Bowes, MapInfo Suite, http://www.pb.com/software/Location-Intelligence/MapInfo-Suite/ (Accessed March 27, 2013).
2. SSTsoftware, SSToolBox, http://www.sstsoftware.com/sstoolbox.htm (Accessed March 27, 2013).
3. María Soledad Escolar Díaz "A generic software architecture for portable applications in heterogeneous wireless sensor networks", Doctoral Thesis 2010, Universidad Carlos III de Madrid. http://e-archivo.uc3m.es/bitstream/10016/9188/1/tesis-SoledadEscolar-Marzo2010.pdf (Accessed March 28, 2013).

4. OGC, Sensor Web Enablement (SWE), http://www.opengeospatial.org/ogc/markets-technologies/swe (Accessed March 27, 2013).

5. SSTsoftware, FarmRite, http://www.sstsoftware.com/farmrite.htm (Accessed March 27, 2013).

6. Agvance, Agvance Mapping, http://www.agvance.net/mapping.html (Accessed March 27, 2013).

7. Agricon, More than technology: Precision Farming is agriculture, http://www.agricon.de/en/products/ (Accessed March 27, 2013).

8. FarmWorks, Information management systems, http://www.farmworks.com/products/office (Accessed March 27, 2013).

9. OGC, OGC® Standards and Supporting Documents, http://www.opengeospatial.org/standards (Accessed March 27, 2013).

10. OGC, SensorML, http://www.ogcnetwork.net/SensorML (Accessed March 27, 2013).

11. QuantumGIS, http://www.qgis.org/ (Accessed March 27, 2013).

12. OGC, Glossary of Terms - C, http://www.opengeospatial.org/ogc/glossary/c (Accessed March 27, 2013).

13. Infrastructure for Spatial Information in the European Community (INSPIRE), http://inspire.jrc.ec.europa.eu/ (Accessed March 27, 2013).

14. GIGAS, http://www.thegigasforum.eu (Accessed March 27, 2013).

15. EuroGEOSS, http://www.eurogeoss.eu (Accessed March 27, 2013).

Chapter 4
Concluding Remarks

Within this text an overview of the state-of-the-art in terms of applications of information and communication technology (ICT) for irrigation management has been provided. The applications of ICT to precision agriculture and emerging field of precision irrigation management draw mainly from three domains: wireless sensor networks, remote sensing and geographic information systems. Wireless sensor networks and remote sensing provide data that can be used to assess the water needs of crops and predict precipitation and soil moisture dynamics in the near future, while the GISs are needed to store data and present it to the shareholders.

Wireless sensor networks and their application to precision agriculture are described as a technology that promises to deliver unparalleled spatial and temporal resolution, when monitoring of agricultural crops is concerned. Different sensors can be used, as well as different communication protocols. This report provides a survey of the major solutions available and discusses their advantages and shortcomings.

Remote sensing in agriculture is a field that traditionally relies on the use of satellite images to assess the health and the needs of crops. When irrigation is concerned satellite imagery can be used to derive myriad geophysical and bio-physical parameters, which can be useful in monitoring a large part of the water cycle, as well as crop health and water requirements at any given point in time. The overview of available and upcoming satellite products, provided in this report, is intended as a succinct repository of knowledge regarding the state-of-the-art satellite products that can be of interest for irrigation management. In addition to comparing different products in terms of their potential, availability and precision, special care was given to state the ways in which a practitioner can obtain this data, as well as the format in which one can expect to get it. While higher spatial and temporal resolution is, naturally, favorable, these requirements can be relaxed when the data is combined with that acquired by the WSNs. On the other hand satellite data can and should be used to derive the optimal placement and architecture of the WSN. While there is clear potential in combining the two technologies, this subject should be explored further in terms of irrigation management.

D. Ćulibrk et al., *Sensing Technologies For Precision Irrigation*,
SpringerBriefs in Electrical and Computer Engineering,
DOI: 10.1007/978-1-4614-8329-8_4, © The Author(s) 2014

GIS is a mature field with numerous commercial and open-source solutions available. The report discusses those most commonly used and provides an overview of their functionality. Although the technology has been around for quite some time, systems intended for precision agriculture are a novel development and none exist which specialize in precision irrigation management.

When designing or purchasing precision agriculture systems, one should be mindful of initiatives such as GEOSS, EuroGEOSS and the INSPIRE directive of the EU, which deal with ways to share spatial information and promise wider applicability of systems that support them. The integration will also enable use of data shared by other agencies to aid irrigation management. The precision irrigation systems should strive towards compatibility and interoperability with as wide a number of other solutions.